Moving Boundary PDE Analysis

Biomedical Applications in R

Moving Boundary PDE Analysis

Biomedical Applications in R

William E. Schiesser

CRC Press
Taylor & Francis Group
Boca Raton London New York

CRC Press is an imprint of the
Taylor & Francis Group, an **informa** business

CRC Press
Taylor & Francis Group
6000 Broken Sound Parkway NW, Suite 300
Boca Raton, FL 33487-2742

First issued in paperback 2023

© 2019 by Taylor & Francis Group, LLC
CRC Press is an imprint of Taylor & Francis Group, an Informa business

No claim to original U.S. Government works

ISBN 13: 978-0-367-22483-7 (hbk)
ISBN-13: 978-1-032-65400-3 (pbk)
ISBN 13: 978-0-429-27512-8 (ebk)

DOI: 10.1201/9780429275128

Library of Congress Cataloging-in-Publication Data

Names: Schiesser, W. E., author.
Title: Moving boundary PDE analysis : biomedical applications in R / by William E. Schiesser.
Description: Boca Raton : Taylor & Francis, [2019]
Identifiers: LCCN 2019003912 | ISBN 9780367224837 (hardback) | ISBN 9780429275128 (ebook)
Subjects: | MESH: Numerical Analysis, Computer-Assisted | Neoplastic Processes | Atherosclerosis | Models, Theoretical
Classification: LCC RC262 | NLM QZ 203 | DDC 616.99/400285--dc23
LC record available at https://lccn.loc.gov/2019003912

Visit the Taylor & Francis Web site at
http://www.taylorandfrancis.com

and the CRC Press Web site at
http://www.crcpress.com

Contents

Preface

This book is directed to the numerical integration (solution) of systems of partial differential equations (PDEs) for which the boundary conditions move in space. In applications, the physical boundaries move as the solution evolves in time.

Moving boundary PDEs (MBPDEs) are an important class of mathematical models with a spectrum of applications. In this book, the focus is on two biomedical applications. The first example (Chapter 4) pertains to the boundary of a tumor that moves outward as the tumor grows. The second example (Chapter 5) pertains to the movement of the inner wall of an artery as plaque forms during atherosclerosis; if the plaque formation continues long enough, the arterial bloodstream is seriously restricted, a precursor to stroke and myocardial infarction (heart attack).

Chapters 1 through 3 discuss a numerical algorithm (computational method) for MFPDEs that to the author's knowledge is an original contribution. The algorithm is implemented in computer routines that are coded in R, a quality open-source scientific programming system. The routines are validated with a series of special-case MFPDE test problems.

The routines are available from a download link so that the reader/analyst/researcher can execute the test problems and example applications discussed in the book without having to first study numerical methods and computer coding. The routines can then be applied to variations and extensions of the reported test problems and applications, such as changes in the MBPDE parameters (constants) and form of the model equations. Finally, the generic routines can be applied to new models, and the associated R routines can be readily executed on modest computers.

I would welcome comments about the application and usefulness of the MBPDE methodology.

W. E. Schiesser
Bethlehem, PA

1

PDE Model Formulation

The intent of this chapter is to introduce moving boundary PDE (MBPDE) analysis through an example based on the linear diffusion equation. In coordinate-free form, the diffusion equation is:

$$\frac{\partial u}{\partial t} = D\nabla^2 u, \tag{1.1a}$$

with u dependent variable, t initial value independent variable, typically time, ∇^2 Laplacian differential operator, D diffusivity.

As a specific example of eq. (1.1a), the Laplacian in cylindrical coordinates (r, θ, z) is:

$$\frac{\partial u}{\partial t} = D\left(\frac{\partial^2 u}{\partial r^2} + \frac{1}{r}\frac{\partial u}{\partial r} + \frac{1}{r^2}\frac{\partial^2 u}{\partial \theta^2} + \frac{\partial^2 u}{\partial z^2}\right). \tag{1.1b}$$

If the derivatives in θ, z are neglected, eq. (1.1b) reduces to the 1D diffusion equation:

$$\frac{\partial u}{\partial t} = D\left(\frac{\partial^2 u}{\partial r^2} + \frac{1}{r}\frac{\partial u}{\partial r}\right). \tag{1.2}$$

Equation (1.2) is first order in t and second order in r, and therefore requires one initial condition (IC) and two boundary conditions (BCs).

$$u(r, t = 0) = u_0(r), \tag{1.3}$$

where $u_0(r)$ is a function to be specified.

$$\frac{\partial u(r = r_l, t)}{\partial r} = 0; \ D\frac{\partial u(r = r_u, t)}{\partial r} = k_m(u_a - u(r = r_u, t)), \tag{1.4a,b}$$

where k_m is a mass transfer coefficient, u_a is a specified ambient value, and r_l, r_u are the lower and upper boundary values of r.

r_u moves according to an equation of motion that usually defines the moving boundary velocity, v_{mb}.

$$\frac{dr_u}{dt} = v_{mb}. \tag{1.5}$$

As an example, the equation of motion of the moving boundary could be of the form:

$$\rho_m \frac{dr_u}{dt} = \pm k_{mb}(u_a - u(r = r_u, t))$$

or

$$\frac{dr_u}{dt} = \pm(k_{mb}/\rho_m)(u_a - u(r = r_u, t)) = \pm k_{ru}(u_a - u(r = r_u, t)), \tag{1.6}$$

where ρ_m is a density (with the units of $u(r, t)$), k_{mb} is a coefficient (with units m/s), and $k_{ru} = (k_{mb}/\rho_m)$.[1]

Equations (1.2) through (1.6) constitute the introductory MBPDE model that is programmed in R[2] as explained in Chapter 2.

---

[1]Equation (1.6) is one possibility for the moving boundary velocity. The numerical algorithm that follows in Chapter 2 can, in principle, accommodate any mathematical form of v_{mb} in eq. (1.5). In the case of eq. (1.6), the choice of a sign in $\pm$ is determined by the sign of $(u_a - u(r = r_u, t))$, so that if, for example, the boundary movement is expected to increase r_u, then $\frac{dr_u}{dt} > 0$.

[2]R is a quality open source scientific programming system that can be easily downloaded from the Internet (http://www.R-project.org/). In particular, R has: (i) vector-matrix operations that facilitate the programming of linear algebra, (ii) a library of quality ordinary differential equation (ODE) integrators, and (iii) graphical utilities for the presentation numerical ODE/PDE solutions. All of these features and utilities are demonstrated through the applications in this book.

2

Model Implementation

2.1 Introduction

The moving boundary partial differential equation (MBPDE) model of eqs. (1.2) through (1.6) is implemented in R routines discussed next. The basic numerical algorithm is the numerical method of lines (MOL) in which the PDE, eq. (1.2), is approximated as a system of initial value ordinary differential equations (ODEs) that can be integrated (solved) numerically with a library ODE integrator [1, 2].

2.2 Programming of MBPDE Model

The following main program is based on the ODE integrator library ODE. The default integrator in ODE is lsoda that automatically switches between algorithms for nonstiff and stiff ODEs [2].

2.2.1 Main Program

The main program for the MOL solution of eqs. (1.2) through (1.6) follows. The steps in the MOL algorithm are explained subsequently, including the corresponding R coding.

Listing 2.1: Main program for eqs. (1.2) through (1.6)

```
#
# 1D, cylindrical coordinates
#
# Delete previous workspaces
  rm(list=ls(all=TRUE))
```

(Continued)

Listing 2.1 (Continued): Main program for eqs. (1.2) through (1.6)

```
#
# Access ODE integrator
  library("deSolve");
#
# Access functions for numerical solution
  setwd("f:/mbpde");
  source("pde_1a.R");
#
# Select case
  ncase=1;
#
# Parameters
  D=1;km=0;ua=1;
  kru=3;
#
# Grid (in r)
  n=21;rl=0;ru=1;
  r=seq(from=rl,to=ru,by=(ru-rl)/(n-1));
#
# Independent variable for ODE integration
  t0=0;
  tout=rep(0,2);
  dt=0.04;np=6;ip=1;
  tp=rep(0,np);rup=rep(0,np);
  drdtp=rep(0,(np-1));
#
# IC for first step in t
  u0=rep(0,n);
  u=rep(0,nrow=n);
  up=matrix(0,nrow=n,ncol=np);
  for(ir in 1:n){
    u0[ir]=exp(-10*r[ir]^2);
     u[ir]=u0[ir];
    up[ir,1]=u[ir];
  }
  tp[1]=t0;rup[1]=ru;
  ncall=0;
```

(*Continued*)

Listing 2.1 (Continued): Main program for eqs. (1.2) through (1.6)

```
#
# Display IC
  cat(sprintf("\n\n       t       r     u(r,t)"));
  iv=seq(from=1,to=n,by=5);
  for(ir in iv){
    cat(sprintf("\n %6.2f%6.2f%10.4f",
        t0,r[ir],u[ir]));
  }
#
# Next step along solution
  while(ip<np){
  for(ir in 1:n){
    u0[ir]=u[ir];
  }
  t0=tout[2];
  tout[1]=t0;
  tout[2]=tout[1]+dt;
#
# ODE integration
  out=ode(func=pde_1a,y=u0,times=tout);
#
# Array for solution
  for(ir in 1:n){
    u[ir]=out[2,ir+1];
  }
#
# Redefine spatial grid
  table=splinefun(r,u);
  if(ncase==1){drdt=0;}
  if(ncase==2){drdt=2;}
  if(ncase==3){drdt=kru*(ua-u[n]);}
  ru=ru+drdt*dt;
  r=seq(from=rl,to=ru,by=(ru-rl)/(n-1));
#
# Solution on redefined grid
  u=table(r,deriv=0);
```

(Continued)

Listing 2.1 (Continued): Main program for eqs. (1.2)
through (1.6)

```
  ip=ip+1;
  rup[ip]=ru;drdtp[ip-1]=drdt;
#
# Display numerical solution
  cat(sprintf("\n\n      t     r     u(r,t)"));
  iv=seq(from=1,to=n,by=5);
  for(ir in iv){
    cat(sprintf("\n %6.2f%6.2f%10.4f",
        tout[2],r[ir],u[ir]));
  }
#
# Solution for plotting
  for(ir in 1:n){
    up[ir,ip]=u[ir];
  }
  tp[ip]=tout[2];
#
# Next step (from while)
  }
#
# Plot output
#
# PDE solution
  matplot(r,up,type="l",lwd=2,col="black",lty=1,
    xlab="r",ylab="u(r,t)");
#
# Boundary position
  plot(tp,rup,xlab="t",ylab="ru");
   lines(tp,rup,type="l",lwd=2);
  points(tp,rup,pch="o",lwd=2);
  plot(tp[2:np],drdtp,xlab="t",ylab="dru/dt");
   lines(tp[2:np],drdtp,type="l",lwd=2);
  points(tp[2:np],drdtp,pch="o",lwd=2);
#
# Calls to ODE routine
  cat(sprintf("\n\n  ncall = %3d\n",ncall));
```

We can note the following details about Listing 2.1.

- Previous workspaces are deleted.

```
#
# 1D, cylindrical coordinates
#
# Delete previous workspaces
  rm(list=ls(all=TRUE))
```

- The R ODE integrator library `deSolve` is accessed. Then the directory with the files for the solution of eqs. (1.2) through (1.6) is designated. Note that `setwd` (set working directory) uses / rather than the usual \.

```
#
# Access ODE integrator
  library("deSolve");
#
# Access functions for numerical solution
  setwd("f:/mbpde");
  source("pde_1a.R");
```

`pde_1a.R` is the routine for the method of lines (MOL) approximation of PDE (1.2) (discussed subsequently in Chapter 3).

- `ncase` is specified with three possible values, `ncase=1,2,3`, corresponding to different right hand sides (RHSs) of eq. (1.6) (different velocities for the moving boundary at $r = r_u$).

```
#
# Select case
  ncase=1;
```

- The model parameters are specified, specifically, D in eqs. (1.2), (1.4b), k_m in eq. (1.4b), k_{ru} in eq. (1.6), and u_a in eqs. (1.4b), (1.6).

```
#
# Parameters
  D=1;km=0;ua=1;
  kru=3;
```

- A spatial grid of 21 points is defined for $x_l = 0 \le x \le x_u = 1$ so that $x = 0, 0.05, ..., 1$.

```
#
# Grid (in r)
  n=21;rl=0;ru=1;
  r=seq(from=rl,to=ru,by=(ru-rl)/(n-1));
```

r_u = ru, the upper limit on r, is changed (increased) for $t > t_0 = 0$
with ncase=2,3, reflecting the moving outer boundary for the MBPDE.

- Parameters for the MOL solution are defined.

```
#
# Independent variable for ODE integration
  t0=0;
  tout=rep(0,2);
  dt=0.04;np=6;ip=1;
  tp=rep(0,np);rup=rep(0,np);
  drdtp=rep(0,(np-1));
```

These statements require some additional explanation.

- The initial value of t for the solution is defined.

    ```
    t0=0
    ```

- Rather than call ode once to compute a complete solution from
  $t_0 = 0$ to a final time $t_f = 0.2$, ode is called for a series of output
  points. In each of the intervals of two points, the grid in r is defined for an updated r_u (calculated by the integration of $\dfrac{dr_u}{dt}$ from
  eq. [1.6]). In this way, r_u is refined as the solution proceeds to
  reflect the moving boundary.

    ```
    tout=rep(0,2);
    ```

- Each interval of two points has length dt=0.04 and six output
  points are defined, np=6 (including $t = t_0$). Therefore, the total
  interval in t for the complete solution is $(6 - 1)(0.04) = 0.2$.

    ```
    dt=0.04;np=6;ip=1;
    ```

- The value of t at the six output points is placed in vector tp and the
  corresponding values of r_u are placed in vector rup for plotting. In
  this way, the movement of the boundary at $r = r_u$ can be observed
  graphically.

    ```
    tp=rep(0,np);rup=rep(0,np);
    ```

- Similarly, the varying values of $\dfrac{dr_u}{dt}$ from eq. (1.6) are placed in
  drdtp for plotting. Since this derivative is not available initially at
  $t = t_0$ (but only after eq. (1.6) is used), there are $6 - 1 = 5$ values
  of the derivative.

    ```
    drdtp=rep(0,(np-1));
    ```

- Function $u_0(r)$ in initial condition (IC) (1.3) is defined as a Gaussian function in r centered at $r = r_l = 0$.

```
#
# IC for first step in t
  u0=rep(0,n);
  u=rep(0,nrow=n);
  up=matrix(0,nrow=n,ncol=np);
  for(ir in 1:n){
    u0[ir]=exp(-10*r[ir]^2);
    u[ir]=u0[ir];
    up[ir,1]=u[ir];
  }
  tp[1]=t0;rup[1]=ru;
  ncall=0;
```

The initial condition u0 is also placed in u for further implementation of the algorithm, and up for plotting. $t = t_0 = 0$ is placed in tp and $r = r_u = 1$ is placed in rup for plotting. The counter for the calls to the ODE/MOL routine pde_1a is also initialized.

- The solution at $t = t_0 = 0$ is displayed for every fifth value of the $n = 21$ values of r (using by=5).

```
#
# Display IC
  cat(sprintf("\n\n      t      r      u(r,t)"));
  iv=seq(from=1,to=n,by=5);
  for(ir in iv){
    cat(sprintf("\n %6.2f%6.2f%10.4f",
        t0,r[ir],u[ir]));
  }
```

- Successive intervals of two points in t are implemented with the while. ip=1 covers the interval $0 \leq t \leq 0.04$ with tout[1]=0, tout[2]=0.04. ip=2 covers the interval $0.04 \leq t \leq 0.08$ with tout[1]=0.04, tout[2]= 0.08, and so forth until $t = t_f = 0.2$ is reached.

```
#
# Next step along solution
  while(ip<np){
  for(ir in 1:n){
    u0[ir]=u[ir];
  }
  t0=tout[2];
  tout[1]=t0;
  tout[2]=tout[1]+dt;
```

- The system of $n = 21$ MOL/ODEs is integrated across each two-point interval by the library integrator ODE (available in deSolve [2]). As expected, the inputs to ODE are the ODE function, pde_1a, the IC vector u0, and the vector of output values of t, tout. The length of u0 (21) informs ODE how many ODEs are to be integrated. func,y,times are reserved names.

```
#
# ODE integration
  out=ode(func=pde_1a,y=u0,times=tout);
```

The numerical solution to the ODEs is returned in matrix out. In this case, out has the dimensions $2 \times (n + 1) = 2 \times 21 + 1 = 22$.

The offset $21 + 1$ is required since the first element of each column has the output t, and the $2, ..., n + 1 = 2, ..., 22$ column elements have the 21 ODE solutions.

- The solution of the 21 ODEs returned in out by ODE is placed in array u.

```
#
# Array for solution
  for(ir in 1:n){
    u[ir]=out[2,ir+1];
  }
```

- Equation (1.6) is integrated to redefine r_u.

```
#
# Redefine spatial grid
  table=splinefun(r,u);
  if(ncase==1){drdt=0;}
  if(ncase==2){drdt=2;}
  if(ncase==3){drdt=kru*(ua-u[n]);}
  ru=ru+drdt*dt;
  r=seq(from=rl,to=ru,by=(ru-rl)/(n-1));
```

This code requires some additional explanation.

- The ICs for the next interval in t are defined.

```
  t0=tout[2];u0=u;ip=ip+1;
  table=splinefun(r,u);
```

splinefun is a function in the basic version of R (it does not have to be accessed as a separate function [1]).

- Three cases for the RHS of eq. (1.6) are defined.

```
if(ncase==1){drdt=0;}
if(ncase==2){drdt=2;}
if(ncase==3){drdt=kru*(ua-u[n]);}
```

For ncase=1, $\dfrac{dr_u}{dt} = 0$ so there is no change in r_u. This case is worth considering since if r_u changes, a programming error is indicated.

For ncase=2, r_u changes at a constant rate so that the variation of u_r with t is linear. Again, this case is worth considering since if this response of r_u is not observed, a programming error is indicated.

For ncase=3, eq. (1.6) is programmed (with +), and the resulting variation of r_u with t can be observed.

ncase=1,2,3 are discussed in Chapter 3.

- Eq. (1.6) is integrated with the explicit Euler method.

```
ru=ru+drdt*dt;
```

dt is presumed small enough that the Euler method gives sufficient accuracy in the calculation of r_u. This assumption can be tested by reducing dt (and increasing np).

- The grid in r is redefined for the new r_u. In other words, the moving boundary (value of r_u) is implemented at this point.

```
r=seq(from=rl,to=ru,by=(ru-rl)/(n-1));
```

- A solution u is computed for the redefined grid r. deriv=0 designates the return of the function (rather than its derivatives in r) interpolated/extrapolated by the spline.

```
#
# Solution on redefined grid
  u=table(r,deriv=0);
  ip=ip+1;
  rup[ip]=ru;drdtp[ip-1]=drdt;
```

This step illustrates the use of an important property of the spline, that is, a different independent variable vector r can be defined and used which permits the implementation of the moving boundary.

The current values of r_u and $\dfrac{dr_u}{dt}$ are updated for subsequent plotting. drdtp[ip-1] is used since the vector drdtp does not include a value of the derivative at $t = t_0 = 0$ (discussed previously).

- The solution (returned by ode in out) is displayed for every fifth value of r (with by=5).

```
#
# Display numerical solution
  cat(sprintf("\n\n      t      r      u(r,t)"));
  iv=seq(from=1,to=n,by=5);
  for(ir in iv){
    cat(sprintf("\n %6.2f%6.2f%10.4f",
        tout[2],r[ir],u[ir]));
  }
```

- The solution $u(r,t)$ and t are placed in up and tp for subsequent plotting.

```
#
# Solution for plotting
  for(ir in 1:n){
    up[ir,ip]=u[ir];
  }
  tp[ip]=tout[2];
```

ip was previously incremented for the next pass through the while.

- The next interval of length dt is programmed within the while (for ip<np).

```
#
# Next step (from while)
  }
```

- At the end of the final step in t (completion of the while), the solution of eq. (1.2) is plotted against r (and parametrically in t).

```
#
# Plot output
#
# PDE solution
  matplot(r,up,type="l",lwd=2,col="black",lty=1,
    xlab="r",ylab="u(r,t)");
```

- The movement of r_u and the associated $\dfrac{du_r}{dt}$ are plotted against t.

```
#
# Boundary position
  plot(tp,rup,xlab="t",ylab="ru");
   lines(tp,rup,type="l",lwd=2);
  points(tp,rup,pch="o",lwd=2);
```

```
   plot(tp[2:np],drdtp,xlab="t",ylab="dru/dt");
    lines(tp[2:np],drdtp,type="l",lwd=2);
   points(tp[2:np],drdtp,pch="o",lwd=2);
```

tp[2:np] is used for t since the derivative array drdtp does not include a value at $t = t_0 = 0$. Both points and a solid line are used (as reflected in the plot that follows in Chapter 3).

- The number of calls to the ODE/MOL routine, pde_1a, is displayed as a measure of the computational effort required to compute the complete solution.

```
#
# Calls to ODE routine
   cat(sprintf("\n\n  ncall = %3d\n",ncall));
```

This completes the discussion of the main program in Listing 2.1. The ODE/MOL routine pde_1a called by ODE in the main program is considered next.

2.2.2 ODE/MOL Routine

ODE/MOL routine pde_1a is listed next.

Listing 2.2: ODE/MOL routine pde_1a for eqs. (1.2) through (1.6)

```
  pde_1a=function(t,u,parms){
#
# Function pde_1a computes the t derivative
# vector of u(r,t)
#
# ur
  tabler=splinefun(r,u);
  ur=tabler(r,deriv=1);
#
# BCs
  ur[1]=0;ur[n]=(km/D)*(ua-u[n]);
#
# urr
  tablerr=splinefun(r,ur);
  urr=tablerr(r,deriv=1);
```

(Continued)

Listing 2.2 (Continued): ODE/MOL routine pde_1a for eqs. (1.2) through (1.6)

```
#
# ODE/PDEs
  ut=rep(0,n);
  for(i in 1:n){
    if(i==1){
      ut[i]=2*D*urr[i];
    }else{
      ut[i]=D*(urr[i]+(1/r[i])*ur[i]);
    }
  }
#
# Increment calls to pde_1a
  ncall <<- ncall+1;
#
# Return derivative vector
  return(list(c(ut)));
  }
```

We can note the following details about Listing 2.2.

- The function is defined.

```
    pde_1a=function(t,u,parms){
    #
    # Function pde_1a computes the t derivative
    # vector of u(r,t)
```

  t is the current value of t in eq. (1.2). u the 21-vector of ODE/MOL dependent variables. parm is an argument to pass parameters to pde_1a (unused, but required in the argument list). The arguments must be listed in the order stated to properly interface with ODE called in the main program of Listing 2.1. The derivative vector of the left hand side (LHS) of eq. (1.2) is calculated and returned to ODE as explained subsequently.

- The first partial derivative $\dfrac{\partial u}{\partial r}$ is computed for the most recent spatial grid r (deriv=1 specifies a first derivative).

```
    #
    # ur
      tabler=splinefun(r,u);
      ur=tabler(r,deriv=1);
```

- BCs (1.4) are programmed.

```
#
# BCs
  ur[1]=0;ur[n]=(km/D)*(ua-u[n]);
```

Subscripts `1`,`n` correspond to $r = r_l, r_u$, respectively.

- The second partial derivative $\dfrac{\partial^2 u}{\partial r^2}$ is computed from the first derivative (successive or stagewise differentiation).

```
#
# urr
  tablerr=splinefun(r,ur);
  urr=tablerr(r,deriv=1);
```

- Equation (1.2) is programmed.

```
#
# ODE/PDEs
  ut=rep(0,n);
  for(i in 1:n){
    if(i==1){
      ut[i]=2*D*urr[i];
    }else{
      ut[i]=D*(urr[i]+(1/r[i])*ur[i]);
    }
  }
```

The derivative $\dfrac{\partial u}{\partial t}$ is placed in vector `ut`. For `i=1` corresponding to $r_l = 0$, the term $\dfrac{1}{r}\dfrac{\partial u}{\partial r}$ is indeterminant $(0/0)$ and is resolved with l'Hospital's rule [1]. That is, the radial group in eq. (1.2) at $r = 0$ is:

$$\frac{\partial^2 u}{\partial r^2} + \frac{1}{r}\frac{\partial u}{\partial r} = 2\frac{\partial^2 u}{\partial r^2} \tag{2.1}$$

and is programmed as `2*D*urr[i]`. For `i>1`, the programming follows directly from eq. (1.2), which demonstrates a principal feature of the MOL (the close correspondence of the PDE and MOL programming).

- The counter for the calls to `pde_1a` is incremented and returned to the main program of Listing 2.1 with `<<-`.

```
#
# Increment calls to pde_1a
  ncall <<- ncall+1;
```

- `ut` is returned to `ODE` as a list (required by `ODE`). `c` is the `R` vector utility.

```
#
# Return derivative vector
  return(list(c(ut)));
  }
```

The final } concludes `pde_1a`.

This concludes the discussion of `pde_1a`. The output from the main program of Listing 2.1 and the subordinate routine `pde_1a` of Listing 2.2 is considered in Chapter 3.

2.3   Summary and Conclusions

An introductory MBPDE consisting of eqs. (1.2) through (1.6) is implemented within the MOL format. The movement of the outer boundary is determined by the equation of motion, eqs. (1.5) and (1.6), an ODE that defines the boundary velocity. This equation is applied to three special cases, but can be generalized to essentially any form that defines the boundary velocity. The output from this introductory MBPDE example for `ncase=1,2,3`, including 3D plotting, is considered next in Chapter 3.

An essential feature of the MOL analysis is the use of a spline that can be used to interpolate/extrapolate onto a redefined grid in space for the next step along the solution. The convergence of this step is examined numerically in Chapter 3, including the Euler integration of eq. (1.6). The spline is also used to calculate the first and second boundary value (spatial) derivatives in the PDE.

In summary, a methodology for MBPDE analysis, demonstrated with an example application, is presented that should be generally useful for MBPDE applications.

References

1. Schiesser, W.E. (2017), *Spline Collocation Methods for Partial Differential Equations*, John Wiley & Sons, Hoboken, NJ.

2. Soetaert, K., J. Cash, and F. Mazzia (2012), *Solving Differential Equations in R*, Springer-Verlag, Heidelberg, Germany.

3

Model Output

3.1 Introduction

In Chapter 1, a basic moving boundary partial differential equation (MBPDE) model, eqs. (1.2) through (1.6), is developed to test a MBPDE algorithm and a computer implementation is discussed in Chapter 2. The output from the R routines documented in Chapter 2 is now consider for a series of cases.

3.2 Fixed Boundary Output

For this first case, the parameters in Listing 2.1 are ncase = 1, km=0 corresponding to no movement of the outer boundary at $r = r_u$ and no mass transfer in eq. (1.4b) across this boundary. These parameters are programmed in the main program of Listing 3.1 as:

Listing 3.1: Parameters for eqs. (1.2) through (1.6),
ncase = 1, km=0

```
#
# Select case
  ncase=1;
#
# Parameters
  D=1;km=0;ua=1;
  kru=3;
```

With no boundary movement (ncase = 1) and no boundary mass transfer km=0, the solution of eq. (1.2) with homogeneous Neumann boundary condition (BC) (1.4b) corresponds to $\dfrac{\partial u(r = r_u, t)}{\partial r} = 0$ reflects no mass transfer through the boundaries. Thus, the Gaussian IC redistributes to a constant value of $u(r,t)$ representing conservation of mass with t. This solution is reflected in the output that follows.[1]

TABLE 3.1

Numerical output for Eqs. (1.2) through (1.6), ncase=1, km=0

t	r	u(r,t)
0.00	0.00	1.0000
0.00	0.25	0.5353
0.00	0.50	0.0821
0.00	0.75	0.0036
0.00	1.00	0.0000

t	r	u(r,t)
0.04	0.00	0.3840
0.04	0.25	0.3025
0.04	0.50	0.1471
0.04	0.75	0.0456
0.04	1.00	0.0172

t	r	u(r,t)
0.08	0.00	0.2378
0.08	0.25	0.2057
0.08	0.50	0.1335
0.08	0.75	0.0712
0.08	1.00	0.0483

t	r	u(r,t)
0.12	0.00	0.1737
0.12	0.25	0.1578
0.12	0.50	0.1196
0.12	0.75	0.0843
0.12	1.00	0.0703

t	r	u(r,t)
0.16	0.00	0.1404
0.16	0.25	0.1321

(Continued)

---

[1] The accuracy of the spline approximation of the Gaussian function $f(r) = e^{-10r^2}$ and its first and second derivatives is discussed in Appendix A1 for interpolation over the interval $0 \leq r \leq 1$, and interpolation/extrapolation over the intervals $0 \leq r \leq 1.25$, and $0 \leq r \leq 1.5$. The accuracy is determined by comparing the spline approximations with the analytical Gaussian and its derivatives as a function of the number of grid points, for both direct and successive (stagewise) differentiation.

TABLE 3.1 (*Continued*)
Numerical output for Eqs. (1.2) through
(1.6), ncase=1, km=0

0.16	0.50	0.1110
0.16	0.75	0.0913
0.16	1.00	0.0832
t	r	u(r,t)
0.20	0.00	0.1222
0.20	0.25	0.1178
0.20	0.50	0.1061
0.20	0.75	0.0952
0.20	1.00	0.0904
ncall = 573		

We can note the following details about the numerical output in Table 3.1.

- IC (1.3) ($t = 0$) is verified for the Gaussian function, e.g., $u(r = 0, t = 0) = 1$.
- The output is for $r = 0, 1/(21 - 1) = 0.05, ..., 1$ as programmed in Listing 2.1 (21 values of r at each value of t with every fifth value in r displayed so $r = 0, 0.25, 0.50, 0.75, 1$).
- The output is for $t = 0, 0.2/(6 - 1) = 0.04, ..., 0.2$ as programmed in Listing 2.1 (6 values of t).
- IC (1.3) is smoothed with increasing t to an approximate constant value of 0.11.
- The values of r do not change with t reflecting a fixed boundary at $r = r_u$.
- The computational effort is modest, ncall = 573 so ODE calculates the solution efficiently.

The graphical output is in Figures 3.1-1 through 3.1-3.

Figure 3.1-1 indicates that the solution starts at the Gaussian function IC and approaches a near constant value. The homogeneous Neumann (zero flux) BCs are clear.

Figure 3.1-2 indicates that r_u remains at 1 (no boundary movement).

Figure 3.1-3 indicates that dr_u/dt remains at 0 (no boundary movement).

This solution is confirmed in two ways: (1) in Appendix A1, a conventional method of lines (MOL) solution is computed with splines for comparison with the MBPDE solution from Listings 2.1, 2.2, i.e., the spline solution directly covers the interval $0 \leq t \leq 0.2$ with one call to ordinary differential equation (ODE) integrator ode, and (2) in Appendix A2, a solution with finite differences (FDs) that parallels the spline solution of (1) is computed. In both

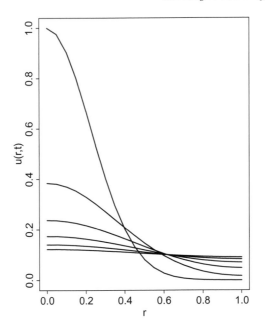

FIGURE 3.1-1
Numerical solution $u(r,t)$ from eq. (1.2), `ncase = 1`, `km=0`.

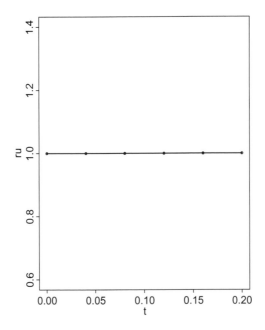

FIGURE 3.1-2
$r_u(t)$ from eq. (1.6), `ncase = 1`, `km=0`.

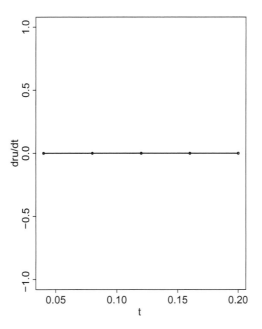

FIGURE 3.1-3
$dr_u(t)/dt$ from eq. (1.6), `ncase = 1`, `km=0`.

cases, the solutions are close to the solution in Table 3.1 and Figure 3.1-1 (see Appendices A1, A2 for the output).

For the next case, mass transfer at $r = r_u$ is included by changing from `km=0` to `km=10` in Listing 2.1 (the second value of `km` was selected to give a pronounced change in the solution). The output (again for `ncase = 1`) follows.

TABLE 3.2
Numerical output for Eqs. (1.2) through
(1.6), `ncase = 1`, `km=10`

t	r	u(r,t)
0.00	0.00	1.0000
0.00	0.25	0.5353
0.00	0.50	0.0821
0.00	0.75	0.0036
0.00	1.00	0.0000
t	r	u(r,t)
0.04	0.00	0.3857
0.04	0.25	0.3107
0.04	0.50	0.2095

(*Continued*)

TABLE 3.2 (*Continued*)
Numerical output for Eqs. (1.2) through
(1.6), ncase = 1, km=10

0.04	0.75	0.3327
0.04	1.00	0.7808

t	r	u(r,t)
0.08	0.00	0.2876
0.08	0.25	0.2913
0.08	0.50	0.3489
0.08	0.75	0.5424
0.08	1.00	0.8619

t	r	u(r,t)
0.12	0.00	0.3306
0.12	0.25	0.3617
0.12	0.50	0.4655
0.12	0.75	0.6551
0.12	1.00	0.8999

t	r	u(r,t)
0.16	0.00	0.4168
0.16	0.25	0.4531
0.16	0.50	0.5597
0.16	0.75	0.7269
0.16	1.00	0.9222

t	r	u(r,t)
0.20	0.00	0.5069
0.20	0.25	0.5409
0.20	0.50	0.6365
0.20	0.75	0.7786
0.20	1.00	0.9375

ncall = 650

The solution in Table 3.2 is clearly substantially different than in Table 3.1. Also, ncall increased to 650.

The differences in the solutions with mass transfer added at $r = r_u$ is indicated by comparing Figures 3.1-1 and 3.2-1.

In Figure 3.2-1, the effect of BC (1.4b) is clear (the solution in the neighborhood of $r = r_u$ approaches $u_a = 1$). Also, the entire solution is increasing and would approach $u(r,t) = u_a$ for a sufficiently large final time (well beyond $t = 0.2$). The plots of $r_u(t)$ and $dr_u(t)/dt$ against t are the same as in Figures 3.1-2 and 3.1-3 and are not repeated here.

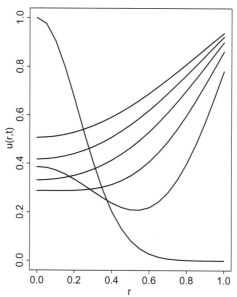

FIGURE 3.2-1
Numerical solution $u(r,t)$ from eq. (1.2), `ncase = 1`, `km=10`.

In summary, the solutions for the preceding cases $(k_m = 0, 10)$, `ncase = 1` are what is expected for a fixed boundary. We next consider the corresponding solutions for `ncase = 2` for which the boundary at $r = r_u$ moves left to right at a specified rate (from Listing 2.1, `if(ncase==2)drdt=2;`).

3.3 Constant Rate Moving Boundary Output

The parameters for `ncase = 2`, `km=0` specfied in the main program of Listing 3.2 are

---

Listing 3.2: Parameters for eqs. (1.2) through (1.6),
ncase = 2, km=0

```
#
# Select case
  ncase=2;
#
# Parameters
  D=1;km=0;ua=1;
  kru=3;
```

The numerical solution is in Table 3.3.

TABLE 3.3

Numerical output for Eqs. (1.2) through
(1.5), ncase = 2, km=0

t	r	u(r,t)
0.00	0.00	1.0000
0.00	0.25	0.5353
0.00	0.50	0.0821
0.00	0.75	0.0036
0.00	1.00	0.0000

t	r	u(r,t)
0.04	0.00	0.3840
0.04	0.27	0.2906
0.04	0.54	0.1255
0.04	0.81	0.0330
0.04	1.08	0.0161

t	r	u(r,t)
0.08	0.00	0.2382
0.08	0.29	0.1955
0.08	0.58	0.1095
0.08	0.87	0.0489
0.08	1.16	0.0375

t	r	u(r,t)
0.12	0.00	0.1737
0.12	0.31	0.1486
0.12	0.62	0.0961
0.12	0.93	0.0567
0.12	1.24	0.0488

t	r	u(r,t)
0.16	0.00	0.1388
0.16	0.33	0.1223
0.16	0.66	0.0874
0.16	0.99	0.0609
0.16	1.32	0.0553

t	r	u(r,t)
0.20	0.00	0.1178
0.20	0.35	0.1063
0.20	0.70	0.0819
0.20	1.05	0.0633
0.20	1.40	0.0592

ncall = 546

We can note the following details about the numerical output in Table 3.3.

- IC (1.3) ($t = 0$) is verified for the Gaussian function, e.g., $u(r = 0, t = 0) = 1$ (as in Tables 3.1 and 3.2).
- The output is for a variable right boundary at $r = r_u$. The boundary movement is according to the Euler integration of eq. (1.6)[2] as summarized below (with `drdt=2`, `dt=0.04`).

$$r_u(1.04) = r_u(0) + \frac{dr_u(0)}{dt} = 1 + (2)(0.04) = 1.08$$

$$r_u(1.08) = r_u(0.04) + \frac{dr_u(0.04)}{dt} = 1.08 + (2)(0.04) = 1.16$$

$$r_u(1.12) = r_u(1.08) + \frac{dr_u(0.08)}{dt} = 1.16 + (2)(0.04) = 1.24$$

$$r_u(0.16) = r_u(0.12) + \frac{dr_u(0.12)}{dt} = 1.24 + (2)(0.04) = 1.32$$

$$r_u(0.20) = r_u(0.16) + \frac{dr_u(0.16)}{dt} = 1.32 + (2)(0.04) = 1.40$$

This variation in $r_u(t)$ is reflected in Table 3.3 and Figures 3.3-1 through 3.3-3.

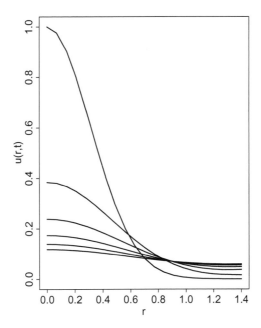

FIGURE 3.3-1

$u(r, t)$ from eq. (1.2), `ncase = 2, km=0`.

---

[2]The Euler method is exact for constant derivatives such as $\frac{dr_u(t)}{dt} = 2$ for `ncase = 2`. For higher order derivatives, the Euler method is an approximation with the accuracy dependent linearly on the integration step $h = dt$, that is, of order $O(h)$.

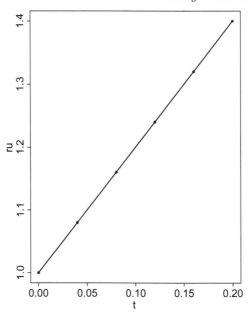

FIGURE 3.3-2
$r_u(t)$ from eq. (1.5), `ncase = 2`, `km=0`.

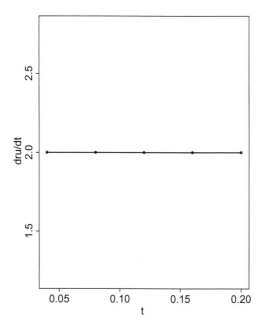

FIGURE 3.3-3
$dr_u(t)/dt$ from eq. (1.5), `ncase = 2`, `km=0`.

- The output is for $t = 0, 0.2/(6-1) = 0.04, ..., 0.20$ as programmed in Listing 2.1 (6 values of t).

- The computational effort is modest, ncall = 546, so ODE calculates the solution efficiently.

The graphical output is in Figures 3.3-1 through 3.3-3.

Figure 3.3-1 indicates that the boundary at r_u moves from $r_u = 1$ to $r_u = 1.4$ as explained by the preceding calculations.

Figure 3.3-2 indicates that r_u varies linearly with a slope of 2.

Figure 3.3-3 indicates that dr_u/dt remains at 2 (Table 3.4).

A second case with a moving boundary (ncase = 2) is illustrated by changing $k_m = 0$ to $k_m = 10$ (in Listing 3.1).

Figure 3.4-1 indicates that the boundary at r_u moves from $r_u = 1$ to $r_u = 1.4$. The plots of $r_u(t)$ and $du_r(t)/dt$ are the same as in Figures 3.3-2

TABLE 3.4

Numerical output for Eqs. (1.2) through (1.6), ncase = 2, km=10

t	r	u(r,t)
0.00	0.00	1.0000
0.00	0.25	0.5353
0.00	0.50	0.0821
0.00	0.75	0.0036
0.00	1.00	0.0000
t	r	u(r,t)
0.04	0.00	0.3857
0.04	0.27	0.3003
0.04	0.54	0.2082
0.04	0.81	0.4143
0.04	1.08	0.9561
t	r	u(r,t)
0.08	0.00	0.2877
0.08	0.29	0.2936
0.08	0.58	0.3847
0.08	0.87	0.6398
0.08	1.16	0.9733
t	r	u(r,t)
0.12	0.00	0.3268
0.12	0.31	0.3694
0.12	0.62	0.5137

(*Continued*)

TABLE 3.4 (*Continued*)
Numerical output for Eqs. (1.2) through
(1.6), ncase = 2, km=10

t	r	u(r,t)
0.12	0.93	0.7457
0.12	1.24	0.9830
t	r	u(r,t)
0.16	0.00	0.4020
0.16	0.33	0.4549
0.16	0.66	0.6053
0.16	0.99	0.8068
0.16	1.32	0.9880
t	r	u(r,t)
0.20	0.00	0.4776
0.20	0.35	0.5310
0.20	0.70	0.6727
0.20	1.05	0.8467
0.20	1.40	0.9911

ncall = 606

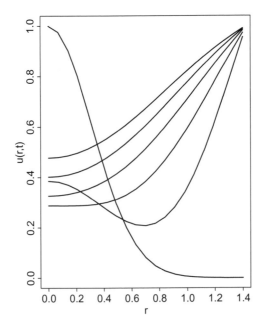

FIGURE 3.4-1
$u(r,t)$ from eq. (1.2), ncase = 2, km=10.

and 3.3-3 and are not repeated here. Changing $k_m = 0$ (Figure 3.3-1) to $k_m = 10$ (Figure 3.4-1) has a substantial effect on the solution.

3.4 Variable Rate Moving Boundary Output

The next case is based on a boundary moving at a variable rate, $r_u(t)$, according to eq. (1.6) (ncase = 3 in Listing 3.1).
   The parameter values in Listing 3.1 are:

```
#
# Select case
   ncase=3;
#
# Parameters
   D=1;km=0;ua=1;
   kru=3;
```

In particular, the rate constant k_{ru} in if(ncase==3)drdt=kru*(ua-u[n]); is set numerically. This value was selected to give a substantial contribution to the position of the moving boundary, $r = r_u$.
   The numerical output is in Table 3.5.
   We can note the following details about this output.

- The Gaussian IC (1.3) is confirmed ($t = 0$).
- The right hand boundary moves over the interval $1 \leq r_u \leq 1.58$
- The computational effort is modest, ncall = 443.

The graphical output is in Figures 3.5-1 through 3.5-3.
   Figure 3.5-1 indicates that the boundary at r_u moves from $r_u = 1$ to $r_u = 1.58$.
   Figure 3.5-2 indicates that r_u varies nearly linearly so that the Euler integration of eq. (1.6) should be accurate.
   Figure 3.5-3 indicates that dr_u/dt decreases modestly with t which accounts for the nearly linear variation of r_u in Figure 3.5-2.
   As the concluding case, the parameters are (with ncase = 3, km=10, kru=3).

```
#
# Select case
   ncase=3;
#
# Parameters
   D=1;km=10;ua=1;
   kru=3;
```

   The numerical output is in Table 3.6.
   We can note the following details about this output.

TABLE 3.5
Numerical output for Eqs. (1.2) through
(1.6), ncase = 3, km=0, kru=3

t	r	u(r,t)
0.00	0.00	1.0000
0.00	0.25	0.5353
0.00	0.50	0.0821
0.00	0.75	0.0036
0.00	1.00	0.0000

t	r	u(r,t)
0.04	0.00	0.3840
0.04	0.28	0.2848
0.04	0.56	0.1159
0.04	0.84	0.0286
0.04	1.12	0.0151

t	r	u(r,t)
0.08	0.00	0.2381
0.08	0.31	0.1904
0.08	0.62	0.0991
0.08	0.93	0.0411
0.08	1.23	0.0334

t	r	u(r,t)
0.12	0.00	0.1736
0.12	0.34	0.1442
0.12	0.67	0.0859
0.12	1.01	0.0471
0.12	1.35	0.0417

t	r	u(r,t)
0.16	0.00	0.1383
0.16	0.37	0.1179
0.16	0.73	0.0774
0.16	1.10	0.0501
0.16	1.46	0.0459

t	r	u(r,t)
0.20	0.00	0.1168
0.20	0.39	0.1017
0.20	0.79	0.0718
0.20	1.18	0.0518
0.20	1.58	0.0484

ncall = 443

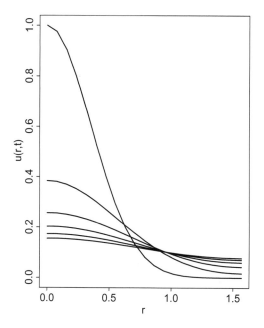

FIGURE 3.5-1
$u(r,t)$ from eq. (1.2), ncase = 3, km=0, kru=3.

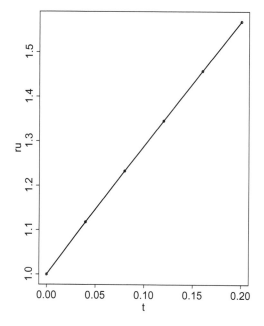

FIGURE 3.5-2
$r_u(t)$ from eq. (1.6), ncase = 3, km=0, kru=3.

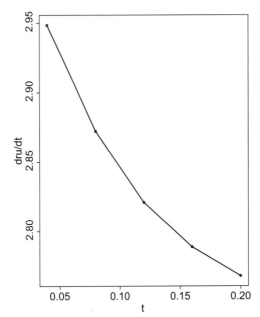

FIGURE 3.5-3
$dr_u(t)/dt$ from eq. (1.6), `ncase = 3`, `km=0`, `kru=3`.

- The Gaussian IC for eq. (1.3) is confirmed $(t = 0)$.
- The variation in r_u has been reduced from the preceding case (Table 3.5) to $1 \leq r_u \leq 1.069$ (the format for the values of r_u was changed from `6.2f` to `7.3f`).
- The solution $u(r,t)$ is increased from the preceding case, and the movement of the boundary $r = r_u$ is reduced substantially.

Table 3.5

t	r	u(r,t)
0.20	0.00	0.1168
0.20	0.39	0.1017
0.20	0.79	0.0718
0.20	1.18	0.0518
0.20	1.58	0.0484

Table 3.6

t	r	u(r,t)
0.20	0.000	0.4951
0.20	0.267	0.5315
0.20	0.535	0.6335
0.20	0.802	0.7824
0.20	1.069	0.9448

TABLE 3.6

Numerical output for Eqs. (1.2) through (1.6), ncase = 3, km=10, kru=3

t	r	u(r,t)
0.00	0.000	1.0000
0.00	0.250	0.5353
0.00	0.500	0.0821
0.00	0.750	0.0036
0.00	1.000	0.0000

t	r	u(r,t)
0.04	0.000	0.3857
0.04	0.257	0.3073
0.04	0.513	0.2083
0.04	0.770	0.3573
0.04	1.026	0.8385

t	r	u(r,t)
0.08	0.000	0.2875
0.08	0.261	0.2916
0.08	0.521	0.3562
0.08	0.782	0.5639
0.08	1.042	0.8887

t	r	u(r,t)
0.12	0.000	0.3286
0.12	0.263	0.3615
0.12	0.527	0.4717
0.12	0.790	0.6695
0.12	1.053	0.9160

t	r	u(r,t)
0.16	0.000	0.4102
0.16	0.266	0.4488
0.16	0.531	0.5615
0.16	0.797	0.7355
0.16	1.062	0.9329

t	r	u(r,t)
0.20	0.000	0.4951
0.20	0.267	0.5315
0.20	0.535	0.6335
0.20	0.802	0.7824
0.20	1.069	0.9448

ncall = 614

This increase in $u(r,t)$ is due to the increased contribution of BC (1.4b) in changing k_m from 0 to 10. The larger values of $u(r,t)$ in turn reduce the difference ua-u[n] in if(ncase==3)drdt=kru*(ua-u[n]); (u[n] is larger and ua remains at ua=1), so that drdt is reduced and therefore r_u from 1.58 to 1.069).

The graphical output is in Figures 3.6-1 through 3.6-3.

Figure 3.6-1 indicates that the boundary at r_u moves from $r_u = 1$ to $r_u = 1.069$.

Figure 3.6-2 indicates that r_u varies with t at a diminishing rate (slope) over $1 \leq r_u \leq 1.072$.

Figure 3.6-3 indicates that dr_u/dt decreases substantially with t which accounts for the variation of r_u in Figure 3.6-2.

3D plotting of the solution $u(r,t)$ can be accomplished by a call to the R utility persp, which is added to the end of Listing 3.1.

```
#
# Calls to ODE routine
  cat(sprintf("\n\n  ncall = %3d\n",ncall));
#
# u(r,t)
  persp(r,tp,up,theta=45,phi=45,
    xlab="r",ylab="t",zlab="u(r,t)");
```

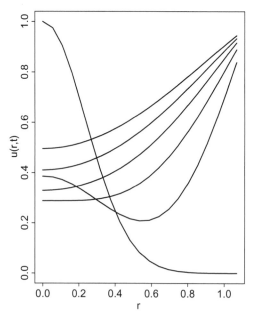

FIGURE 3.6-1

$u(r,t)$ from eq. (1.2), ncase = 3, km=10, kru=3.

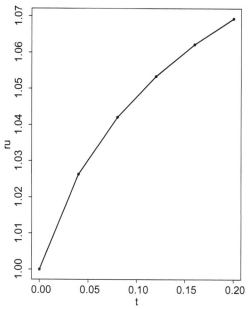

FIGURE 3.6-2
$r_u(t)$ from eqs. (1.5) and (1.6), ncase = 3, km=10, kru=3.

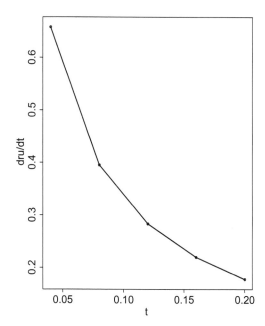

FIGURE 3.6-3
$dr_u(t)/dt$ from eqs. (1.5) and (1.6), ncase = 3, km=10, kru=3.

If the solution $u(r, t)$ is to be plotted in 3D (persp) in addition to 2D (from matplot in Listing 3.1) such as in Figure 3.6-1, additional curves are required in t (beyond 6 from np=6). The following changes give 21 solution curves rather than 6.

```
dt=0.04;np=6;ip=1;
```

changed to

```
dt=0.01;np=21;ip=1;
```

This decrease in dt will also test the convergence in t of the Euler integration of eq. (1.6).

Abbreviated output for this case, with the previous parameters for Figures 3.6-1 through 3.6-3, follows in Table 3.7.

We can note the following details about this output.

- The Gaussian IC of eq. (1.3) is confirmed $(t = 0)$.
- The output is for dt=0.01.
- The final value of r_u is 1.083. Therefore, the change in the final value with the increase in np is:

  ```
  r_u=1.069, dt=0.04, np=6
  ```

  ```
  r_u=1.083, dt=0.01, np=21
  ```

  or $(1.083 - 1.069)/1.083 \times 100 = 1.3\%$.

- The computational effort is acceptable, ncall = 1064.

The graphical output is in Figures 3.7-1 through 3.7-4.

Figure 3.7-1 indicates that the boundary at r_u moves from $r_u = 1$ to $r_u = 1.083$ and 21 curves parametric in t.

Figure 3.7-2 indicates that r_u varies with t at a decreasing rate (slope) over $1 \leq r_u \leq 1.083$.

Figure 3.7-3 indicates that dr_u/dt decreases substantially with t which accounts for the variation of r_u in Figure 3.7-2.

Figure 3.7-4 indicates that the resolution in t is adequate for the 3D plot from persp.

This change in np (from 6 to 21) can be extended in the same way to test the 3D plotting and convergence of the Euler integration (Table 3.8). For example, with:

```
dt=0.005;np=41;ip=1;
```

the output is (the format for t is changed from 6.2f to 7.3f):

TABLE 3.7
Numerical output for Eqs. (1.2) through
(1.6) ncase = 3, km=10, kru=3, dt=0.01,
np=21

t	r	u(r,t)
0.00	0.000	1.0000
0.00	0.250	0.5353
0.00	0.500	0.0821
0.00	0.750	0.0036
0.00	1.000	0.0000

t	r	u(r,t)
0.01	0.000	0.7129
0.01	0.253	0.4522
0.01	0.506	0.1147
0.01	0.759	0.0514
0.01	1.012	0.6460

.
.
.

Output for t = 0.02 to
0.18 removed

.
.
.

t	r	u(r,t)
0.19	0.000	0.4641
0.19	0.270	0.5019
0.19	0.540	0.6083
0.19	0.811	0.7661
0.19	1.081	0.9390

t	r	u(r,t)
0.20	0.000	0.4844
0.20	0.271	0.5214
0.20	0.541	0.6249
0.20	0.812	0.7767
0.20	1.083	0.9418

ncall = 1064

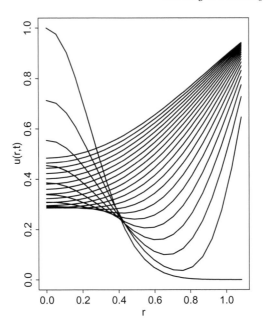

FIGURE 3.7-1
$u(r,t)$ from eq. (1.2), ncase = 3, km=10, kru=3, dt=0.01, np=21.

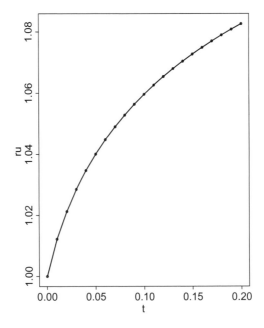

FIGURE 3.7-2
$r_u(t)$ from eq. (1.6), ncase = 3, km=10, kru=3, dt=0.01, np=21.

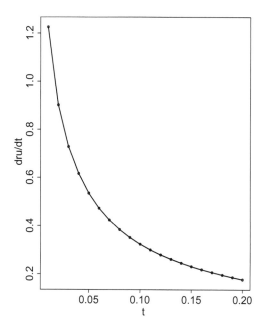

FIGURE 3.7-3
$dr_u(t)/dt$ from eq. (1.5), ncase = 3, km=10, kru=3, dt=0.01, np=21.

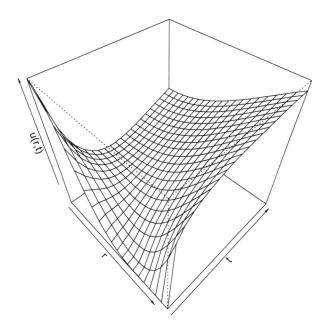

FIGURE 3.7-4
$u(r, t)$ from eq. (1.2), ncase = 3, km=10, kru=3 dt=0.01, np=21.

TABLE 3.8

Numerical output for Eqs. (1.2) through
(1.6), ncase = 3, km=10, kru=3, dt=0.005,
np=41.

t	r	u(r,t)
0.000	0.000	1.0000
0.000	0.250	0.5353
0.000	0.500	0.0821
0.000	0.750	0.0036
0.000	1.000	0.0000

t	r	u(r,t)
0.005	0.000	0.8320
0.005	0.252	0.4912
0.005	0.504	0.1004
0.005	0.756	0.0120
0.005	1.008	0.5403

.   .
.   .
.   .

Output for t = 0.010 to
0.190 removed

.   .
.   .
.   .

t	r	u(r,t)
0.195	0.000	0.4715
0.195	0.271	0.5090
0.195	0.542	0.6144
0.195	0.814	0.7700
0.195	1.085	0.9399

t	r	u(r,t)
0.200	0.000	0.4816
0.200	0.271	0.5187
0.200	0.543	0.6226
0.200	0.814	0.7752
0.200	1.086	0.9413

ncall = 1384

We can note the following details about this output.

- The Gaussian IC of eq. (1.3) is confirmed ($t = 0$).
- The output is for `dt=0.005`.
- The final value of r_u is `1.086`. Therefore, the change in the final value with the increase in `np` is:

  `r_u=1.083, dt=0.01, np=21`

  `r_u=1.086, dt=0.005, np=41`

  or $(1.086 - 1.083)/1.083 \times 100 = 0.28\%$.

  This measure of convergence is based on small departures of r_u from 1. A more stringent test of convergence would follow from the larger increases in r_u as reflected in Table 3.5 ($r_u = 1.58$ for `np=6`), by using `np=21,41`. This test is left as an exercise.

- The computational effort increases, `ncall` = `1384`, as expected with additional output points.

The graphical output is in Figures 3.8-1 through 3.8-4.

Figure 3.8-1 indicates that the boundary at r_u moves from $r_u = 1$ to $r_u = 1.086$ and 41 curves parametric in t.

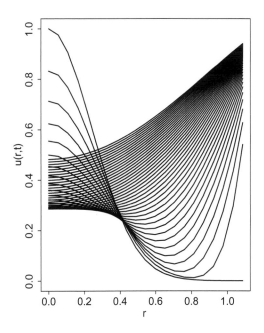

FIGURE 3.8-1

$u(r, t)$ from eq. (1.2), ncase = 3, km=10, kru=3, dt=0.005, np=41.

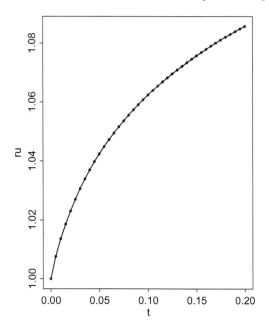

FIGURE 3.8-2
$r_u(t)$ from eq. (1.6), `ncase = 3`, `km=10`, `kru=3`, `dt=0.005`, `np=41`.

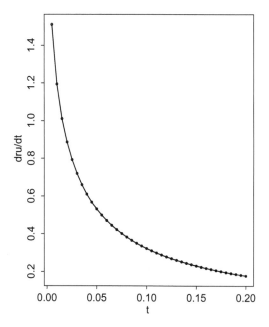

FIGURE 3.8-3
$dr_u(t)/dt$ from eq. (1.6), `ncase = 3`, `km=10`, `kru=3`, `dt=0.005`, `np=41`.

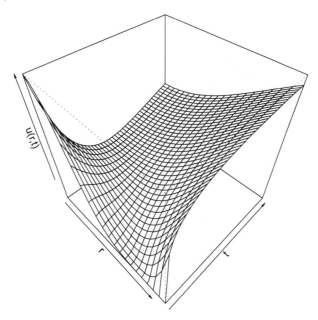

FIGURE 3.8-4

$u(r,t)$ from eq. (1.2), `ncase = 3`, `km=10`, `kru=3`, `dt=0.005`, `np=41`.

Figure 3.8-2 indicates that r_u varies with t at a diminishing rate (slope) over $1 \le r_u \le 1.086$.

Figure 3.8-3 indicates that dr_u/dt decreases substantially with t which accounts for the variation of r_u in Figure 3.8-2.

Figure 3.8-4 indicates that the resolution in t and x is acceptable for the 3D plot from `persp`.

The convergence in t has been verified numerically. The convergence in r can also be studied by increasing `n=21` in Listing 3.1; this is left as an exercise.

3.5 Summary and Conclusions

The MOL solution of eqs. (1.2) through (1.6) implemented in the R routines in Chapter 2 has been studied in this chapter for `ncase = 1,2,3`. This analysis indicates that as `km, kru` in BC (1.4b) and eq. (1.6) are varied, the solutions are quite complicated, so that numerical methods are required (analytical methods are probably precluded). However, a numerical methodology is now available for MBPDE applications as considered in subsequent chapters.

4

Tumor Growth

4.1 Introduction

The moving boundary partial differential equation (MBPDE) algorithm developed in Chapters 1 and 2 is now applied to biomedical applications, starting with a MBPDE model for the growth of a cancer tumor. The model consists of a set of three PDEs as explained next.

4.2 PDEs in Coordinate-free Format

The three PDEs for the tumor growth model are first stated in coordinate-free format [2] (the notation is listed in Table 4.1).

$$\frac{\partial u_1}{\partial t} = \nabla^2 u_1 - \chi_1 \nabla \cdot (u_1 \nabla u_2) - \chi_2 \nabla \cdot (u_1 \nabla u_3) + \mu u_1 (r_1 - u_1 - u_3) \quad (4.1a)$$

$$\sigma \frac{\partial u_2}{\partial t} = \nabla^2 u_2 + r_2(-u_2 + u_1) \quad (4.1b)$$

$$\frac{\partial u_3}{\partial t} = -r_3 u_2 u_3 + \eta u_3 (1 - u_3 - u_2) \quad (4.1c)$$

The physical interpretation of the various left and right hand side (LHS and RHS) terms in eq. (4.1) is discussed subsequently. Briefly, chemotaxis pertains to the movement of the cells in response to the gradient of the MDE. Haptotaxis pertains to the movement of the cells in response to a gradient in ECM.

Equation (4.1) is now specialized to spherical coordinates, (r, θ, ϕ) which are considered a better fit to the geometry of a tumor than 2D and 3D Cartesian coordinates used in [2].

TABLE 4.1
Variables, functions, operators, parameters in eq. (4.1)

Variable, function operator, parameter	Interpretation
u_1	density of cancer cells
u_2	concentration of the matrix degrading enzyme (MDE)
u_3	surrounding extra-cellular matrix (ECM)
t	time
$\nabla\cdot$	divergence of a vector
∇	gradient of a scalar
χ_1	chemotactic sensitivity
χ_2	haptotactic sensitivity
μ, σ, η	parameters (constants)
r_1, r_2, r_3	

4.3 Vector Operators in Spherical Coordinates

From [1, 3], the spatial differential operators (denoted with ∇) in spherical coordinates are given in Tables 4.2 and 4.3.

Symmetry with respect to the angles (θ, ϕ), is assumed so that only derivatives with respect to r appear in the PDEs. The advantage of spherical coordinates is that the model is reduced to 1D.

TABLE 4.2
$\nabla\cdot$ (Divergence of a vector, spherical coordinates)

$$
\begin{bmatrix}
[\nabla]_r = \dfrac{1}{r^2}\dfrac{\partial}{\partial r}(r^2) \\[2ex]
[\nabla]_\theta = \dfrac{1}{r\sin\theta}\dfrac{\partial}{\partial\theta}(\sin\theta) \\[2ex]
[\nabla]_\phi = \dfrac{1}{r\sin\theta}\dfrac{\partial}{\partial\phi}
\end{bmatrix}
$$

TABLE 4.3
∇ (Gradient of a scalar, spherical coordinates)

$$
\begin{bmatrix}
[\nabla]_r = \dfrac{\partial}{\partial r} \\[2mm]
[\nabla]_\theta = \dfrac{1}{r}\dfrac{\partial}{\partial \theta} \\[2mm]
[\nabla]_\phi = \dfrac{1}{r \sin \theta}\dfrac{\partial}{\partial \phi}
\end{bmatrix}
$$

4.4 1D PDE Model, Spherical Coordinates

With angular variations neglected, eq. (4.1) becomes (with the vector differential operators $\nabla\cdot, \nabla$ from Tables 4.2 and 4.3):

$$
\frac{\partial u_1}{\partial t} = \frac{1}{r^2}\frac{\partial}{\partial r}\left(r^2 \frac{\partial u_1}{\partial r}\right)
$$

$$
-\chi_1 \frac{1}{r^2}\frac{\partial}{\partial r}\left(r^2 u_1 \frac{\partial u_2}{\partial r}\right) - \chi_2 \frac{1}{r^2}\frac{\partial}{\partial r}\left(r^2 u_1 \frac{\partial u_3}{\partial r}\right) + \mu u_1(r_1 - u_1 - u_3) \quad (4.2a)
$$

$$
\sigma \frac{\partial u_2}{\partial t} = \frac{1}{r^2}\frac{\partial}{\partial r}\left(r^2 \frac{\partial u_2}{\partial r}\right) + r_2(-u_2 + u_1) \quad (4.2b)
$$

$$
\frac{\partial u_3}{\partial t} = -r_3 u_2 u_3 + \eta u_3(1 - u_3 - u_2). \quad (4.2c)
$$

Expanding the radial groups:

$$
\frac{\partial u_1}{\partial t} = \frac{\partial^2 u_1}{\partial r^2} + \frac{2}{r}\frac{\partial u_1}{\partial r}
$$

$$
-\chi_1 \left(u_1 \frac{\partial^2 u_2}{\partial r^2} + \frac{\partial u_1}{\partial r}\frac{\partial u_2}{\partial r} + \frac{2}{r}u_1 \frac{\partial u_2}{\partial r} \right)
$$

$$
-\chi_2 \left(u_1 \frac{\partial^2 u_3}{\partial r^2} + \frac{\partial u_1}{\partial r}\frac{\partial u_3}{\partial r} + \frac{2}{r}u_1 \frac{\partial u_3}{\partial r} \right) + \mu u_1(r_1 - u_1 - u_3) \quad (4.3a)
$$

$$
\sigma \frac{\partial u_2}{\partial t} = \frac{\partial^2 u_2}{\partial r^2} + \frac{2}{r}\frac{\partial u_2}{\partial r} + r_2(-u_2 + u_1) \quad (4.3b)
$$

$$\frac{\partial u_3}{\partial t} = -r_3 u_2 u_3 + \eta u_3 (1 - u_3 - u_2). \tag{4.3c}$$

For $r = 0$, eq. (4.3) becomes[1]:

$$\frac{\partial u_1}{\partial t} = 3 \frac{\partial^2 u_1}{\partial r^2}$$

$$- \chi_1 \left(3 u_1 \frac{\partial^2 u_2}{\partial r^2} + \frac{\partial u_1}{\partial r} \frac{\partial u_2}{\partial r} \right)$$

$$- \chi_2 \left(3 u_1 \frac{\partial^2 u_3}{\partial r^2} + \frac{\partial u_1}{\partial r} \frac{\partial u_3}{\partial r} \right) + \mu u_1 (r_1 - u_1 - u_3) \tag{4.4a}$$

$$\sigma \frac{\partial u_2}{\partial t} = 3 \frac{\partial^2 u_2}{\partial r^2} + r_2(-u_2 + u_1) \tag{4.4b}$$

$$\frac{\partial u_3}{\partial t} = -r_3 u_2 u_3 + \eta u_3 (1 - u_3 - u_2). \tag{4.4c}$$

Equations (4.3a, 4.3b) and (4.4a, 4.4b) are second order in r, and they therefore each require two boundary conditions (BCs).

$$\frac{\partial u_1(r = 0, t)}{\partial r} = 0 \tag{4.5a}$$

$$\frac{\partial u_1(r = r_u, t)}{\partial r}$$

$$- \chi_1 u_1(r = r_u, t) \frac{\partial u_2(r = r_u, t)}{\partial r}$$

$$- \chi_2 u_1(r = r_u, t) \frac{\partial u_3(r = r_u, t)}{\partial r} = 0 \tag{4.5b}$$

$$\frac{\partial u_2(r = 0, t)}{\partial r} = \frac{\partial u_2(r = r_u, t)}{\partial r} = 0 \tag{4.5c,d}$$

Equation (4.5) are termed *homogeneous, Neumann* BCs. They are Neumann since they specify the derivative of the solutions at the boundaries.[2] They are homogeneous since the RHSs are zero. Physically, eq. (4.5) specify no diffusion or flux of the cells or MDE through the boundaries ($r = r_l = 0, r_u$). They are therefore termed *no-flux* or *impermeable* BCs. In the case of eq. (4.5a)

---

[1] At $r = 0$, the first-order radial groups are indeterminate. Application of l'Hospital's rule ([5], p298) gives $\frac{2}{r} \frac{\partial u_1}{\partial r} = 2 \frac{\partial^2 u_1}{\partial r^2}$, $\frac{2}{r} \frac{\partial u_2}{\partial r} = 2 \frac{\partial^2 u_2}{\partial r^2}$, $\frac{2}{r} \frac{\partial u_3}{\partial r} = 2 \frac{\partial^2 u_3}{\partial r^2}$.

[2] BCs that specify the PDE-dependent variables at the boundaries are termed *Dirichlet*.

and (4.5c) the zero derivatives result from *symmetry* of the solutions around $r = r_l = 0$. BC (4.5b) expresses zero flux for the cancer cells at $r = r_u$ including linear, chemotaxis, and haptotaxis diffusion. Equations (4.3c) and (4.4c) do not require BCs since they do not have derivatives in r.

These BCs demonstrate the advantage of using spherical coordinates rather than Cartesian coordinates, that is, the BCs are naturally and conveniently expressed in spherical coordinates.

Equations (4.3) and (4.4) are first order in t, and they therefore each require one initial condition (IC).

$$u_1(r, t = 0) = h_1(r); \quad u_2(r, t = 0) = h_2(r); \quad u_3(r, t = 0) = h_3(r). \quad (4.6a,b,c)$$

h_1, h_2, h_3 are functions to be specified.

We conclude this section with a brief explanation (interpretation) of the terms in eq. (4.2), which are component (mass) balances[3] on a differential shell of thickness dr and volume $4\pi r^2 dr$.

For eq. (4.3a):

- $\dfrac{\partial^2 u_1}{\partial r^2} + \dfrac{2}{r} \dfrac{\partial u_1}{\partial r}$: Conventional Fick's second law for the net diffusion of cells into or out of the differential volume $4\pi r^2 dr$.

- $-\chi_1 \left(u_1 \dfrac{\partial^2 u_2}{\partial r^2} + \dfrac{\partial u_1}{\partial r} \dfrac{\partial u_2}{\partial r} + \dfrac{2}{r} u_1 \dfrac{\partial u_2}{\partial r} \right)$: Chemotaxis diffusion into or out of the differential volume $4\pi r^2 dr$. The diffusion flux is given by:

$$\chi_1 u_1 \dfrac{\partial u_2}{\partial r}$$

which is a departure from the conventional Fick's first law with the following features:

1. The rate of diffusion of the cells is proportional to the gradient of MDE, $\dfrac{\partial u_2}{\partial r}$, rather than the gradient of the cells, $\dfrac{\partial u_1}{\partial r}$.
2. The sign of the flux is opposite to that of Fick's first law, so the cells move in the direction of *increasing* MDE concentration, u_2. Thus, MDE is termed an *attractant* and the term models *chemotaxis* (movement in response to MDE).
3. The rate of diffusion is also proportional to the concentration of cells, u_1. The product $u_1 \dfrac{\partial u_2}{\partial r}$ is therefore nonlinear (since it depends on the two dependent variables u_1, u_2).
4. The parameter χ_1 serves as a type of diffusivity.

---

[3]A detailed derivation of mass balances in Cartesian, cylindrical and spherical coordinates is given in [4], Appendix 1.

- $-\chi_2 \left(u_1 \dfrac{\partial^2 u_3}{\partial r^2} + \dfrac{\partial u_1}{\partial r} \dfrac{\partial u_3}{\partial r} + \dfrac{2}{r} u_1 \dfrac{\partial u_3}{\partial r} \right)$: Haptotaxis diffusion into or out of the differential volume $4\pi r^2 dr$ in response to ECM, u_3. The diffusion flux is given by:

$$\chi_2 u_1 \frac{\partial u_3}{\partial r}$$

which has the features:

1. The rate of diffusion of the cells is proportional to the gradient of ECM, $\dfrac{\partial u_3}{\partial r}$, rather than the gradient of the cells, $\dfrac{\partial u_1}{\partial r}$.

2. The sign of the flux is opposite to that of Fick's first law, so the cells move in the direction of *increasing* ECM, u_3. Thus, the ECM is an *attractant*, and this term acts in the same way as the first term for chemotaxis.

3. The rate of diffusion is also proportional to the concentration of cells, u_1. The product $u_1 \dfrac{\partial u_3}{\partial r}$ is therefore nonlinear (since it depends on the two dependent variables u_1, u_3).

4. The parameter χ_2 serves as a type of diffusivity.

- The net chemotaxis-haptotaxis depends on the additive effect of the two terms (with χ_1 and χ_2).

- $+\mu u_1 (r_1 - u_1 - u_3)$: A volumetric logistic (nonlinear) source term for the cells.

- $\dfrac{\partial u_1}{\partial t}$: The accumulation (when positive) or depletion (when negative) of cells in the differential volume $4\pi r^2 dr$. The variation with t depends on the additive effect of the conventional, chemotaxis and haptotaxis diffusion of the cells, and the volumetric source term, as expressed by the RHS of eq. (4.3a).

In summary, eq. (4.3a) is a nonlinear diffusion equation for which a numerical MOL algorithm is programmed in the routines discussed next.

For eq. (4.3b):

- $\dfrac{\partial^2 u_2}{\partial r^2} + \dfrac{2}{r} \dfrac{\partial u_2}{\partial r}$: Conventional Fick's second law for the net diffusion of MDE into or out the differential volume $4\pi r^2 dr$.

- $r_2(-u_2 + u_1)$: Volumetric rate of consumption of MDE and production of MDE from the cells.

- $\sigma \dfrac{\partial u_2}{\partial t}$: The accumulation (when positive) or depletion (when negative) of MDE in the differential volume $4\pi r^2 dr$. The variation with t depends on the additive effect of the conventional diffusion of the MDE, and the volumetric source terms, as expressed by the RHS of eq. (4.3b). σ is a time constant that determines the rate of change of u_2.

In summary, eq. (4.3b) is a linear diffusion equation programmed in the routines discussed next.

For eq. (4.3c):

- $-r_3 u_2 u_3$: Consumption rate through the combined effect of u_2 and u_3.

- $\eta u_3 (1 - u_3 - u_2)$: Volumetric logistic source term.

- $\dfrac{\partial u_3}{\partial t}$: The accumulation (when positive) or depletion (when negative) of ECM in the differential volume $4\pi r^2 dr$. The variation with t depends on the combined effect of the consumption rate term and the logistic source term, as expressed by the RHS of eq. (4.3c).

Equation (4.3c) is a PDE even though a derivative in r does not appear in the RHS (u_3 is a function of r and t through u_2 in $-r_3 u_2 u_3$ and through u_1 in $\eta(1 - u_3 - u_1)$).

The MOL implementation of eqs. (4.3) through (4.6) in a series of R routines is considered next.

4.4.1 Main Program

The following main program is an extension of the main program in Listing 2.1 for three PDEs rather than one.

Listing 4.1: Main program for eqs. (4.3) through (4.6)

```
#
# Tumor growth
#
# Delete previous workspaces
  rm(list=ls(all=TRUE))
#
# Access ODE integrator
  library("deSolve");
#
# Access functions for numerical solution
  setwd("f:/mbpde/chap4");
  source("pde_1a.R");
  source("dss004.R");
  source("dss044.R");
```

(Continued)

Listing 4.1 (Continued): Main program for eqs. (4.3)
through (4.6)

```
#
# Select case for moving boundary
  ncase=1;
#
# Parameters
  npara=1;
#
# Chemotaxis-haptotaxis not included
# (note chi1, chi2)
  if(npara==1){
    d1=0.5e-08;d2=0.5e-08;
    chi1=0.0e-08;chi2=0.0e-08;
    eta=1.0e-08;mu=1.0e-07;
    r1=1;r2=1.0e-07;r3=3.0e-06;
    kru=3.0e-06;
  }
#
# Chemotaxis-haptotaxis included
  if(npara==2){
    d1=0.5e-08;d2=0.5e-08;
    chi1=1.0e-08;chi2=1.0e-08;
    eta=1.0e-08;mu=1.0e-07;
    r1=1;r2=1.0e-07;r3=3.0e-06;
    kru=3.0e-06;
  }
#
# Grid (in r)
  n=21;rl=0;ru=1;
  r=seq(from=rl,to=ru,by=(ru-rl)/(n-1));
#
# Independent variable for ODE integration
  t0=0;
  tout=rep(0,2);
  dt=2.0e+06;np=21;ip=1;
  tp=rep(0,np);rup=rep(0,np);
  drdtp=rep(0,(np-1));
```

                                                    (*Continued*)

Listing 4.1 (Continued): Main program for eqs. (4.3) through (4.6)

```
#
# IC for first step in t
  u0=rep(0,3*n);
  u1=rep(0,n);
  u2=rep(0,n);
  u3=rep(0,n);
 u1p=matrix(0,nrow=n,ncol=np);
 u2p=matrix(0,nrow=n,ncol=np);
 u3p=matrix(0,nrow=n,ncol=np);
  for(i in 1:n){
    u0[i]=exp(-10*r[i]^2);
    u1[i]=u0[i];u1p[i,1]=u1[i];
    u0[i+n]=0;
    u2[i]=u0[i+1];u2p[i,1]=u2[i];
    u0[i+2*n]=1-u1[i];
    u3[i]=u0[i+2*n];u3p[i,1]=u3[i];
  }
  tp[1]=t0;rup[1]=ru;
  ncall=0;
#
# Display IC
  cat(sprintf("\n\n            t      r    u1(r,t)"));
  cat(sprintf("\n          t      r    u2(r,t)"));
  cat(sprintf("\n          t      r    u3(r,t)"));
  iv=seq(from=1,to=n,by=5);
  for(i in iv){
    cat(sprintf("\n %8.2e%6.2f%10.4f",
        t0,r[i],u1[i]));
    cat(sprintf("\n %8.2e%6.2f%10.4f",
        t0,r[i],u2[i]));
    cat(sprintf("\n %8.2e%6.2f%10.4f\n",
        t0,r[i],u3[i]));
  }
#
# Next step along solution
  while(ip<np){
```

(*Continued*)

Listing 4.1 (Continued): Main program for eqs. (4.3) through (4.6)

```
  for(i in 1:n){
        u0[i]=u1[i];
      u0[i+n]=u2[i];
    u0[i+2*n]=u3[i];
  }
  t0=tout[2];
  tout[1]=t0;
  tout[2]=tout[1]+dt;
#
# ODE integration
  out=ode(func=pde_1a,y=u0,times=tout);
#
# Arrays for solution
  for(i in 1:n){
    u1[i]=out[2,i+1];
    u2[i]=out[2,i+1+n];
    u3[i]=out[2,i+1+2*n];
  }
#
# Redefine spatial grid
  table1=splinefun(r,u1);
  table2=splinefun(r,u2);
  table3=splinefun(r,u3);
  if(ncase==1){drdt=0;}
  if(ncase==2){drdt=1.0e-08;}
  if(ncase==3){drdt=kru*u1[n]*(1-u1[n]-u3[n])*u3[n];}
  ru=ru+drdt*dt;
  r=seq(from=rl,to=ru,by=(ru-rl)/(n-1));
#
# Solution on redefined grid
  u1=table1(r,deriv=0);
  u2=table2(r,deriv=0);
  u3=table3(r,deriv=0);
  ip=ip+1;
  rup[ip]=ru;drdtp[ip-1]=drdt;
#
# Display numerical solution
  if((ip-1)*(ip-11)*(ip-21)==0){
  cat(sprintf("\n\n          t       r    u1(r,t)"));
  cat(sprintf("\n        t       r    u2(r,t)"));
```

(*Continued*)

Listing 4.1 (Continued): Main program for eqs. (4.3) through (4.6)

```
  cat(sprintf("\n         t      r    u3(r,t)"));
  iv=seq(from=1,to=n,by=5);
  for(i in iv){
    cat(sprintf("\n %8.2e%6.2f%10.4f",
        tout[2],r[i],u1[i]));
    cat(sprintf("\n %8.2e%6.2f%10.4f",
        tout[2],r[i],u2[i]));
    cat(sprintf("\n %8.2e%6.2f%10.4f\n",
        tout[2],r[i],u3[i]));
  }
  }
#
# Solution for plotting
  for(i in 1:n){
    u1p[i,ip]=u1[i];
    u2p[i,ip]=u2[i];
    u3p[i,ip]=u3[i];
  }
  tp[ip]=tout[2];
#
# Next step (from while)
  }
#
# Plot output
#
# 2D
  matplot(r,u1p,type="l",lwd=2,col="black",lty=1,
    xlab="r",ylab="u1(r,t)");
  matplot(r,u2p,type="l",lwd=2,col="black",lty=1,
    xlab="r",ylab="u2(r,t)");
  matplot(r,u3p,type="l",lwd=2,col="black",lty=1,
    xlab="r",ylab="u3(r,t)");
#
# 3D
  persp(r,tp,u1p,theta=60,phi=45,
        xlim=c(rl,ru),ylim=c(tp[1],tp[np]),xlab="r",
        ylab="t",zlab="u1(r,t)");
```

(Continued)

Listing 4.1 (Continued): Main program for eqs. (4.3) through (4.6)

```
  persp(r,tp,u2p,theta=60,phi=45,
        xlim=c(rl,ru),ylim=c(tp[1],tp[np]),xlab="r",
        ylab="t",zlab="u2(r,t)");
  persp(r,tp,u3p,theta=60,phi=45,
        xlim=c(rl,ru),ylim=c(tp[1],tp[np]),xlab="r",
        ylab="t",zlab="u3(r,t)");
#
# Boundary position
  mon=60*60*24*30;
  plot(tp/mon,rup,xlab="t",ylab="ru");
   lines(tp/mon,rup,type="l",lwd=2);
  points(tp/mon,rup,pch="o",lwd=2);
  plot(tp[2:np]/mon,drdtp,xlab="t",ylab="dru/dt");
   lines(tp[2:np]/mon,drdtp,type="l",lwd=2);
  points(tp[2:np]/mon,drdtp,pch="o",lwd=2);
#
# Calls to ODE routine
  cat(sprintf("\n\n  ncall = %3d\n",ncall));
```

We can note the following details about Listing 4.1.

- Previous workspaces are deleted.

```
  #
  # Tumor growth
  #
  # Delete previous workspaces
    rm(list=ls(all=TRUE))
```

- The R ODE integrator library deSolve is accessed. Then the directory with the files for the solution of eqs. 4.3 through 4.6 is designated. Note that setwd (set working directory) uses / rather than the usual \.

```
  #
  # Access ODE integrator
    library("deSolve");
  #
  # Access functions for numerical solution
    setwd("f:/mbpde/chap4");
    source("pde_1a.R");
    source("dss004.R");
    source("dss044.R");
```

pde_1a.R is the routine for the method of lines (MOL) approximation of PDEs (4.3) and (4.4) (discussed subsequently). dss004, dss044 are library routines for the calculation of first and second order spatial derivatives, respectively.

- ncase is specified with three possible values, ncase=1,2,3, corresponding to different velocities for the moving boundary at $r = r_u$.

```
# Select case for moving boundary
  ncase=1;
```

- The model parameters are specified, with npara=1 for no chemotaxis-haptotaxis nonlinear diffusion and npara=2 for chemotaxis-haptotaxis diffusion included.

```
#
# Parameters
  npara=1;
#
# Chemotaxis-haptotaxis not included
# (note chi1, chi2)
  if(npara==1){
    d1=0.5e-08;d2=0.5e-08;
    chi1=0.0e-08;chi2=0.0e-08;
    eta=1.0e-08;mu=1.0e-07;
    r1=1;r2=1.0e-07;r3=3.0e-06;
    kru=3.0e-06;
  }
#
# Chemotaxis-haptotaxis included
  if(npara==2){
    d1=0.5e-08;d2=0.5e-08;
    chi1=1.0e-08;chi2=1.0e-08;
    eta=1.0e-08;mu=1.0e-07;
    r1=1;r2=1.0e-07;r3=3.0e-06;
    kru=3.0e-06;
  }
```

(based on the selection of χ_1, χ_2 in eqs. (4.3a) and (4.4a)).

- A spatial grid of 21 points is defined for $r_l = 0 \leq r \leq r_u = 1$, so that $r = 0, 0.05, ..., 1$.

```
#
# Grid (in r)
  n=21;rl=0;ru=1;
  r=seq(from=rl,to=ru,by=(ru-rl)/(n-1));
```

$r_u = $ ru, the upper limit on r, is changed (increased), reflecting the moving outer boundary for the MBPDE.

- Parameters for the MOL solution are defined.

```
#
# Independent variable for ODE integration
  t0=0;
  tout=rep(0,2);
  dt=2.0e+06;np=21;ip=1;
  tp=rep(0,np);rup=rep(0,np);
  drdtp=rep(0,(np-1));
```

These statements require some additional explanation.

- The initial value of t for the solution is defined.

```
  t0=0
```

- Rather than call ode once to compute a complete solution from t_0 to a final time t_f, ode is called for a series of output points in vector tout. In each of these intervals of two points, the grid in r is defined for an updated r_u (calculated by ru=ru+drdt*dt). In this way, r_u is refined as the solution proceeds to reflect the moving boundary.

```
  tout=rep(0,2);
```

- Each interval of two points has length dt=2e+06 and 21 output points are defined, np=21 (including $t = t_0$). Therefore, the total interval in t for the complete solution is $(21 - 1)(2.0e + 06) = 4.0e + 07$. The first point in each interval has index ip=1, and the second point has index ip=2.

```
  dt=2.0e+06;np=21;ip=1;
```

- The value of t at the 21 output points is placed in vector tp and the corresponding values of r_u are placed in vector rup for plotting. In this way, the movement of the boundary at $r = r_u$ can be observed graphically.

```
  tp=rep(0,np);rup=rep(0,np);
```

- Similarly, the varying values of $\dfrac{dr_u}{dt}$ (for ncase=1,2,3) are placed in drdtp for plotting. Since this derivative is not available initially at $t = t_0$ (but only after ru=ru+drdt*dt is used), there are $21-1 = 20$ values of the derivative.

```
  drdtp=rep(0,(np-1));
```

- Function $h_1(r)$ in IC (4.6a) is defined as a Gaussian function in r centered at $r = r_l = 0$. $h_2(r)$ in IC (4.6b) is the zero function, and

$h_3(r)$ in IC (4.6c) is $1 - h_1(r)$. This is the beginning of the coding for simultaneous PDEs (eqs. (4.3) and (4.4)).

```
#
# IC for first step in t
  u0=rep(0,3*n);
  u1=rep(0,n);
  u2=rep(0,n);
  u3=rep(0,n);
 u1p=matrix(0,nrow=n,ncol=np);
 u2p=matrix(0,nrow=n,ncol=np);
 u3p=matrix(0,nrow=n,ncol=np);
  for(i in 1:n){
    u0[i]=exp(-10*r[i]^2);
    u1[i]=u0[i];u1p[i,1]=u1[i];
    u0[i+n]=0;
    u2[i]=u0[i+1];u2p[i,1]=u2[i];
    u0[i+2*n]=1-u1[i];
    u3[i]=u0[i+2*n];u3p[i,1]=u3[i];
  }
  tp[1]=t0;rup[1]=ru;
  ncall=0;
```

The three IC functions are placed in a single vector u0 for the start of the integration of 3*n = 3*21 = 63 MOL ODEs. Also, the ICs are placed in u1, u2, u3 for subsequent use in the moving boundary algorithm and in u1p, u2p, u3p for plotting. The initial values $t = $ tp[1] and $r_u = $ rup[1] are also defined. The counter for the calls to the ODE/MOL routine pde_1a is initialized.

• The solutions $u_1(r, t = 0)$, $u_2(r, t = 0)$, $u_3(r, t = 0)$ are displayed for every fifth value of the $n = 21$ values of r (using by=5).

```
#
# Display IC
  cat(sprintf("\n\n          t      r    u1(r,t)"));
  cat(sprintf("\n          t      r    u2(r,t)"));
  cat(sprintf("\n          t      r    u3(r,t)"));
  iv=seq(from=1,to=n,by=5);
  for(i in iv){
    cat(sprintf("\n %8.2e%6.2f%10.4f",
        t0,r[i],u1[i]));
    cat(sprintf("\n %8.2e%6.2f%10.4f",
        t0,r[i],u2[i]));
    cat(sprintf("\n %8.2e%6.2f%10.4f\n",
        t0,r[i],u3[i]));
  }
```

- The next interval of two points is initialized. For `ip=1`, the first interval $0 \le t \le dt = 2.0e + 06$ is initialized (using the `while`).

```
#
# Next step along solution
  while(ip<np){
  for(i in 1:n){
        u0[i]=u1[i];
      u0[i+n]=u2[i];
    u0[i+2*n]=u3[i];
  }
  t0=tout[2];
  tout[1]=t0;
  tout[2]=tout[1]+dt;
```

- The system of $n = 63$ MOL/ODEs is integrated by the library integrator `ODE` (available in `deSolve` [3]). As expected, the inputs to `ODE` are the ODE function, `pde_1a`, the IC vector `u0`, and the vector of output values of t, `tout`. The length of `u0` $(3(21) = 63)$ informs `ODE` how many ODEs are to be integrated. `func,y,times` are reserved names.

```
#
# ODE integration
  out=ode(func=pde_1a,y=u0,times=tout);
```

The numerical solution to the ODEs is returned in matrix `out`. In this case, `out` has the dimensions $2 \times 3n + 1 = 2 \times 63 + 1 = 64$.

The offset $63 + 1$ is required since the first element of each column has the output t (also in `tout`), and the $2, ..., 3n + 1 = 2, ..., 64$ column elements have the 63 ODE solutions.

- The solution of the 63 ODEs returned in `out` by `ODE` is placed in vectors `u1`, `u2`, `u3` of length 21 (defined previously).

```
#
# Arrays for solution
  for(i in 1:n){
    u1[i]=out[2,i+1];
    u2[i]=out[2,i+1+n];
    u3[i]=out[2,i+1+2*n];
  }
```

- `ru=ru+drdt*dt` is used to redefine r_u.

```
#
# Redefine spatial grid
```

```
table1=splinefun(r,u1);
table2=splinefun(r,u2);
table3=splinefun(r,u3);
if(ncase==1){drdt=0;}
if(ncase==2){drdt=1.0e-08;}
if(ncase==3){drdt=kru*u1[n]*(1-u1[n]-u3[n])*u3[n];}
ru=ru+drdt*dt;
r=seq(from=rl,to=ru,by=(ru-rl)/(n-1));
```

This code requires some additional explanation.

- A table of spline coefficients is defined for each dependent variable at the current r with `splinefun`.

  ```
  table1=splinefun(r,u1);
  table2=splinefun(r,u2);
  table3=splinefun(r,u3);
  ```

- Three cases for $\dfrac{dr_u}{dt} = $ `drdt` are defined.

  ```
  if(ncase==1){drdt=0;}
  if(ncase==2){drdt=1.0e-08;}
  if(ncase==3){drdt=kru*u1[n]*(1-u1[n]-u3[n])*u3[n];}
  ```

For `ncase=1`, $\dfrac{dr_u}{dt} = 0$, so there is no change in r_u. This case is worth considering since if r_u changes, a programming error is indicated.

For `ncase=2`, r_u changes at a constant rate `1.0e-08`, so that the variation of r_u with t is linear. Again, this case is worth considering since if this response of r_u is not observed, a programming error is indicated.

For `ncase=3`, `drdt` is defined by the nonlinear logistic function

`kru*u1[n]*(1-u1[n]-u3[n])*u3[n]`

to reflect the combined effect of the cell density and concentration of ECM at the outer boundary in $r = r_u$. This function for `drdt` is suggested by the RHS terms of eqs. (4.3a), (4.4a) (for `u1`), and eqs. (4.3c), (4.4c) (for `u3`), but other choices for the moving boundary function can be programmed here. A particular choice will determine how the tumor outer boundary at $r = r_u$ moves with t.

- The explicit Euler method is used to compute the next r_u.

  ```
  ru=ru+drdt*dt;
  ```

  `dt` is presumed small enough that the Euler method gives sufficient accuracy in the calculation of r_u.

- The grid in r is redefined for the new r_u. In other words, the moving grid is implemented at this point.

  ```
  r=seq(from=rl,to=ru,by=(ru-rl)/(n-1));
  ```

- Modified solutions `u1`,`u2`,`u3` are computed for the redefined grid r. `deriv=0` designates the return of the function (rather than its derivatives in r) interpolated/extrapolated by the spline.

  ```
  #
  # Solution on redefined grid
    u1=table1(r,deriv=0);
    u2=table2(r,deriv=0);
    u3=table3(r,deriv=0);
  ```

  This step illustrates the use of an important property of the spline, that is, a different independent variable vector r can be defined and used which permits the implementation of the moving grid.

- The current values of r_u and $\dfrac{dr_u}{dt}$ are updated for subsequent plotting.

  ```
  ip=ip+1;
  rup[ip]=ru;drdtp[ip-1]=drdt;
  ```

  `drdtp[ip-1]` is used since the vector `drdtp` does not include a value of the derivative at $t = t_0 = 0$ (discussed previously).

- The solution (returned by `ode` in `out`) is displayed for every fifth value of r (with `by=5`) at $t = 0, 2.0 \times 10^7, 4.0 \times 10^7$ (`ip=0,11,21`).

  ```
  #
  # Display numerical solution
    if((ip-1)*(ip-11)*(ip-21)==0){
    cat(sprintf("\n\n          t        r     u1(r,t)"));
    cat(sprintf("\n         t        r     u2(r,t)"));
    cat(sprintf("\n         t        r     u3(r,t)"));
    iv=seq(from=1,to=n,by=5);
    for(i in iv){
      cat(sprintf("\n %8.2e%6.2f%10.4f",
          tout[2],r[i],u1[i]));
  ```

```
        cat(sprintf("\n %8.2e%6.2f%10.4f",
            tout[2],r[i],u2[i]));
        cat(sprintf("\n %8.2e%6.2f%10.4f\n",
            tout[2],r[i],u3[i]));
    }
}
```

- The solutions and t are also placed in arrays for plotting.

```
#
# Solutions for plotting
    for(i in 1:n){
        u1p[i,ip]=u1[i];
        u2p[i,ip]=u2[i];
        u3p[i,ip]=u3[i];
    }
    tp[ip]=tout[2];
```

- The next step in t of length dt is initiated (for ip<np).

```
#
# Next step (from while)
    }
```

- At the end of the final step in t, the solution of eqs. (4.3) and (4.4) is plotted in 2D with matplot.

```
#
# Plot output
#
# 2D
    matplot(r,u1p,type="l",lwd=2,col="black",lty=1,
        xlab="r",ylab="u1(r,t)");
    matplot(r,u2p,type="l",lwd=2,col="black",lty=1,
        xlab="r",ylab="u2(r,t)");
    matplot(r,u3p,type="l",lwd=2,col="black",lty=1,
        xlab="r",ylab="u3(r,t)");
```

- The solutions are also plotted in 3D with persp.

```
#
# 3D
    persp(r,tp,u1p,theta=60,phi=45,
            xlim=c(rl,ru),ylim=c(tp[1],tp[np]),xlab="r",
            ylab="t",zlab="u1(r,t)");
```

```
persp(r,tp,u2p,theta=60,phi=45,
      xlim=c(rl,ru),ylim=c(tp[1],tp[np]),xlab="r",
      ylab="t",zlab="u2(r,t)");
persp(r,tp,u3p,theta=60,phi=45,
      xlim=c(rl,ru),ylim=c(tp[1],tp[np]),xlab="r",
      ylab="t",zlab="u3(r,t)");
```

- The movement of r_u and the associated $\dfrac{dr_u}{dt}$ are plotted against t.

```
#
# Boundary position
  mon=60*60*24*30;
  plot(tp/mon,rup,xlab="t",ylab="ru");
    lines(tp/mon,rup,type="l",lwd=2);
  points(tp/mon,rup,pch="o",lwd=2);
  plot(tp[2:np]/mon,drdtp,xlab="t",ylab="dru/dt");
    lines(tp[2:np]/mon,drdtp,type="l",lwd=2);
  points(tp[2:np]/mon,drdtp,pch="o",lwd=2);
```

`tp[2:np]` is used for t since the derivative array `drdtp` does not include a value at $t = t_0 = 0$. The scaling of t is in months rather than seconds with `mon`.

- The number of calls to the ODE/MOL routine, `pde_1a`, is displayed as a measure of the computational effort required to compute the complete solution.

```
#
# Calls to ODE routine
  cat(sprintf("\n\n  ncall = %3d\n",ncall));
```

This completes the discussion of the main program in Listing 4.1. The ODE/MOL routine `pde_1a` called by `ODE` in the main program is considered next.

4.4.2 ODE/MOL Routine

Finite differences (FD) are used in `pde_1a` for the calculation of spatial derivatives in r (rather than splines as in Listing 2.2) to demonstrate an alternate approach to MOL analysis.

Listing 4.2: ODE/MOL routine for eqs. (4.3) through (4.5)

```
pde_1a=function(t,u,parms){
#
# Function pde_1a computes the t derivative
# vectors of u1(r,t),u2(r,t),u3(r,t)
#
# One vector to three vectors
  u1=rep(0,n);u2=rep(0,n);u3=rep(0,n);
  for(i in 1:n){
    u1[i]=u[i];
    u2[i]=u[i+n];
    u3[i]=u[i+2*n];
  }
#
# u1r,u2r,u3r
  u1r=dss004(rl,ru,n,u1);
  u2r=dss004(rl,ru,n,u2);
  u3r=dss004(rl,ru,n,u3);
#
# BCs
  u1r[1]=0; u1r[n]=chi1*u1[n]*u2r[n]+
                   chi2*u1[n]*u3r[n];
  u2r[1]=0; u2r[n]=0;
#
# u1rr,u2rr,u3rr
  nl=2;nu=2;
  u1rr=dss044(rl,ru,n,u1,u1r,nl,nu);
  u2rr=dss044(rl,ru,n,u2,u2r,nl,nu);
  u3rr=dss044(rl,ru,n,u3,u3r,nl,nu);
#
# PDEs
  u1t=rep(0,n);u2t=rep(0,n);u3t=rep(0,n);
  for(i in 1:n){
    if(i==1){
      u1t[i]=3*d1*u1rr[i]-
             chi1*3*u1[i]*u2rr[i]-
             chi2*3*u1[i]*u3rr[i]+
             mu*u1[i]*(r1-u1[i]-u3[i]);
      u2t[i]=3*d2*u2rr[i]+r2*(-u2[i]+u1[i]);
      u3t[i]=r3*(-u2[i]*u3[i])+eta*u3[i]*
             (1-u3[i]-u1[i]);
    }
```

(Continued)

Listing 4.2 (Continued): ODE/MOL routine for eqs. (4.3) through (4.5)

```
  if(i>1){
    u1t[i]=d1*(u1rr[i]+(2/r[i])*u1r[i])-
        chi1*(u1[i]*u2rr[i]+u1r[i]*u2r[i]+
        (2/r[i])*u1[i]*u2r[i])-
        chi2*(u1[i]*u3rr[i]+u1r[i]*u3r[i]+
        (2/r[i])*u1[i]*u3r[i])+
        mu*u1[i]*(r1-u1[i]-u3[i]));
    u2t[i]=d2*(u2rr[i]+(2/r[i])*u2r[i])+
        r2*(-u2[i]+u1[i]);
    u3t[i]=r3*(-u2[i]*u3[i])+eta*u3[i]*
            (1-u3[i]-u1[i]);
    }
  }
#
# Three vectors to one vector
  ut=rep(0,3*n);
  for(i in 1:n){
    ut[i]    =u1t[i];
    ut[i+n]  =u2t[i];
    ut[i+2*n]=u3t[i];
  }
#
# Increment calls to pde_1a
  ncall <<- ncall+1;
#
# Return derivative vector
  return(list(c(ut)));
  }
```

We can note the following details about Listing 4.2.

- The function is defined.

```
    pde_1a=function(t,u,parms){
  #
  # Function pde_1a computes the t derivative
  # vectors of u1(r,t),u2(r,t),u3(r,t)
```

t is the current value of t in eqs. (4.3) and (4.4). u is the 63-vector of the ODE/MOL dependent variables. **parm** is an argument to pass parameters to **pde_1a** (unused, but required in the argument list). The arguments must be listed in the order stated to properly interface with

ODE called in the main program of Listing 4.1. The derivative vector of the LHS of eqs. (4.3) and (4.4) is calculated and returned to ODE as explained subsequently.

- u is placed in three vectors, u1, u2, u3, to facilitate the programming of eqs. (4.3) and (4.4).

```
#
# One vector to three vectors
  u1=rep(0,n);u2=rep(0,n);u3=rep(0,n);
  for(i in 1:n){
    u1[i]=u[i];
    u2[i]=u[i+n];
    u3[i]=u[i+2*n];
  }
```

- The first partial derivatives $\dfrac{\partial u_1}{\partial r}, \dfrac{\partial u_2}{\partial r}, \dfrac{\partial u_3}{\partial r}$, are computed for the most recent spatial grid r.

```
#
# u1r,u2r,u3r
  u1r=dss004(rl,ru,n,u1);
  u2r=dss004(rl,ru,n,u2);
  u3r=dss004(rl,ru,n,u3);
```

- BCs (4.5) are programmed.

```
#
# BCs
  u1r[1]=0; u1r[n]=chi1*u1[n]*u2r[n]+
                   chi2*u1[n]*u3r[n];
  u2r[1]=0; u2r[n]=0;
```

Subscripts 1,n correspond to $r = r_l, r_u$, respectively.

- The second partial derivatives $\dfrac{\partial^2 u_1}{\partial r^2}, \dfrac{\partial^2 u_2}{\partial r^2}, \dfrac{\partial^2 u_3}{\partial r^2}$, are computed for the most recent spatial grid r, including the BCs.

```
#
# u1rr,u2rr,u3rr
  nl=2;nu=2;
  u1rr=dss044(rl,ru,n,u1,u1r,nl,nu);
  u2rr=dss044(rl,ru,n,u2,u2r,nl,nu);
  u3rr=dss044(rl,ru,n,u3,u3r,nl,nu);
```

nl=2, nu=2 specify Neumann BCs.

- Equations (4.3) and (4.4) are programmed.

```
#
# PDEs
```

```
u1t=rep(0,n);u2t=rep(0,n);u3t=rep(0,n);
for(i in 1:n){
  if(i==1){
    u1t[i]=3*d1*u1rr[i]-chi1*3*u1[i]*u2rr[i]-
                 chi2*3*u1[i]*u3rr[i]+
                 mu*u1[i]*(r1-u1[i]-u3[i]);
    u2t[i]=3*d2*u2rr[i]+r2*(-u2[i]+u1[i]);
    u3t[i]=r3*(-u2[i]*u3[i])+eta*u3[i]*
           (1-u3[i]-u1[i]);
  }
  if(i>1){
    u1t[i]=d1*(u1rr[i]+(2/r[i])*u1r[i])-
      chi1*(u1[i]*u2rr[i]+u1r[i]*u2r[i]+
      (2/r[i])*u1[i]*u2r[i])-
      chi2*(u1[i]*u3rr[i]+u1r[i]*u3r[i]+
      (2/r[i])*u1[i]*u3r[i])+
      mu*u1[i]*(r1-u1[i]-u3[i]);
    u2t[i]=d2*(u2rr[i]+(2/r[i])*u2r[i])+
      r2*(-u2[i]+u1[i]);
    u3t[i]=r3*(-u2[i]*u3[i])+eta*u3[i]*
           (1-u3[i]-u1[i]);
  }
}
```

The derivative $\dfrac{\partial u_1}{\partial t}$ is placed in vector u1t. i=1, is for eq. (4.3a) and i>1 is for eq. (4.4a). Similarly, the derivatives $\dfrac{\partial u_2}{\partial t}$ and $\dfrac{\partial u_3}{\partial t}$ are placed in vectors u2t, u3t. The close correspondence of the PDEs and the MOL programming demonstrates a principal feature of the MOL.

- The counter for the calls to pde_1a is incremented and returned to the main program of Listing 4.1 with <<-.

```
#
# Increment calls to pde_1a
  ncall <<- ncall+1;
```

- ut is returned to ODE as a list (required by ODE). c is the R vector utility.

```
#
# Return derivative vector
  return(list(c(ut)));
}
```

The final } concludes pde_1a.

This concludes the discussion of **pde_1a**. The output from the main program of Listing 4.1 and the subordinate routine **pde_1a** of Listing 4.2 is considered next, starting with **ncase=1**.

4.4.3   Model Output

The numerical output for **ncase=1**, **npara=2** in the main program of Listing 4.1 follows (Table 4.4).

We can note the following details about this output (including the graphical output in Figures 4.1-1 through 4.1-8).

TABLE 4.4
Abbreviated output from the routines in
Listings 4.1 and 4.2, ncase=1, npara=2

t	r	u1(r,t)
t	r	u2(r,t)
t	r	u3(r,t)
0.00e+00	0.00	1.0000
0.00e+00	0.00	0.0000
0.00e+00	0.00	0.0000
0.00e+00	0.25	0.5353
0.00e+00	0.25	0.0000
0.00e+00	0.25	0.4647
0.00e+00	0.50	0.0821
0.00e+00	0.50	0.0000
0.00e+00	0.50	0.9179
0.00e+00	0.75	0.0036
0.00e+00	0.75	0.0000
0.00e+00	0.75	0.9964
0.00e+00	1.00	0.0000
0.00e+00	1.00	0.0000
0.00e+00	1.00	1.0000
t	r	u1(r,t)
t	r	u2(r,t)
t	r	u3(r,t)
2.00e+07	0.00	0.1488
2.00e+07	0.00	0.0916

(*Continued*)

TABLE 4.4 (*Continued*)
Abbreviated output from the routines in
Listings 4.1 and 4.2, ncase=1, npara=2

2.00e+07	0.00	0.0000
2.00e+07	0.25	0.1294
2.00e+07	0.25	0.0827
2.00e+07	0.25	0.0052
2.00e+07	0.50	0.0948
2.00e+07	0.50	0.0649
2.00e+07	0.50	0.0585
2.00e+07	0.75	0.0838
2.00e+07	0.75	0.0530
2.00e+07	0.75	0.2395
2.00e+07	1.00	0.0883
2.00e+07	1.00	0.0499
2.00e+07	1.00	0.3626
t	r	u1(r,t)
t	r	u2(r,t)
t	r	u3(r,t)
4.00e+07	0.00	0.4000
4.00e+07	0.00	0.2356
4.00e+07	0.00	0.0000
4.00e+07	0.25	0.3962
4.00e+07	0.25	0.2337
4.00e+07	0.25	0.0000
4.00e+07	0.50	0.3903
4.00e+07	0.50	0.2306
4.00e+07	0.50	0.0000
4.00e+07	0.75	0.3888
4.00e+07	0.75	0.2293
4.00e+07	0.75	0.0002
4.00e+07	1.00	0.3893
4.00e+07	1.00	0.2293
4.00e+07	1.00	0.0003

ncall = 3851

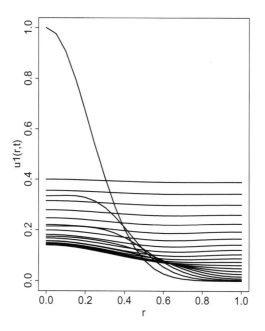

FIGURE 4.1-1
$u_1(r,t)$ from eq. (4.3a), `ncase=1`.

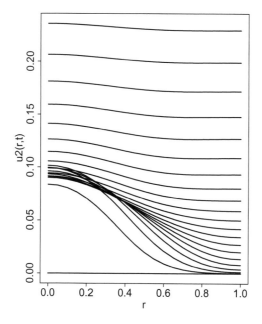

FIGURE 4.1-2
$u_2(r,t)$ from eq. (4.3b), `ncase=1`.

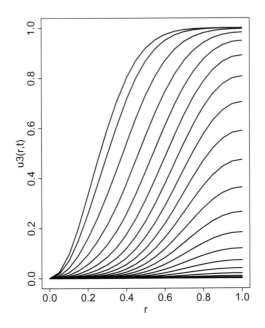

FIGURE 4.1-3
$u_3(r,t)$ from eq. (4.3c), ncase=1.

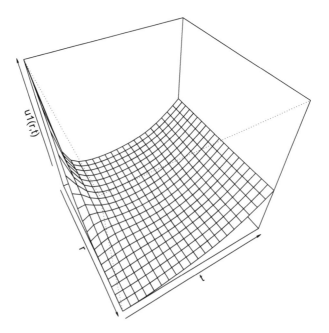

FIGURE 4.1-4
$u_1(r,t)$ from eq. (4.3a), ncase=1.

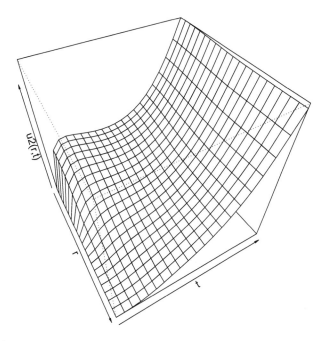

FIGURE 4.1-5
$u_2(r,t)$ from eq. (4.3b), `ncase=1`.

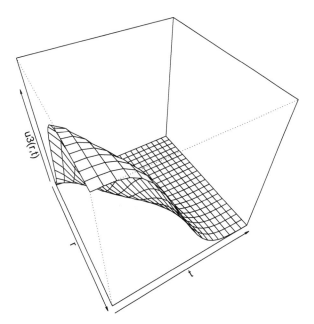

FIGURE 4.1-6
$u_3(r,t)$ from eq. (4.3c), `ncase=1`.

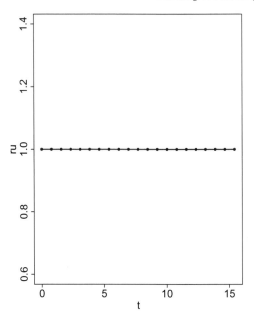

FIGURE 4.1-7
ru from `ru=ru+drdt*dt`, ncase=1.

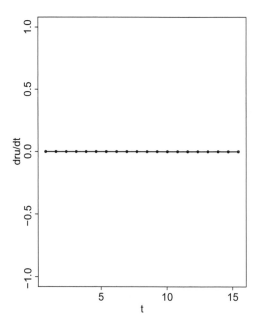

FIGURE 4.1-8
drdt, ncase=1.

- The Gaussian IC for $u_1(r,t)$, $u_1(r,t=0) = e^{-10r^2}$, appears to be confirmed, e.g., $u_1(r=0, t=0) = e^{-10(0)} = 1$ (see Listing 4.1).

```
     t        r      u1(r,t)
 0.00e+00   0.00    1.0000
 0.00e+00   0.25    0.5353
 0.00e+00   0.50    0.0821
 0.00e+00   0.75    0.0036
 0.00e+00   1.00    0.0000
```

Other specific values for $t = 0$ can be verified.

- The homogeneous IC for $u_2(r, t = 0)$ is confirmed.

```
     t        r      u2(r,t)
 0.00e+00   0.00    0.0000
 0.00e+00   0.25    0.0000
 0.00e+00   0.50    0.0000
 0.00e+00   0.75    0.0000
 0.00e+00   1.00    0.0000
```

- The IC $u_3(r, t = 0) = 1 - u_1(r, t = 0)$ is confirmed.

```
     t        r      u3(r,t)
 0.00e+00   0.00    0.0000
 0.00e+00   0.25    0.4647
 0.00e+00   0.50    0.9179
 0.00e+00   0.75    0.9964
 0.00e+00   1.00    1.0000
```

The check of the ICs is important since if they are incorrect, the subsequent solution will be incorrect.

- The values $r = 0, 0.25, 0.5, 0.75, 1$ correspond to 21 values of r over $r_l = 0 \leq r \leq r_u = 1$ with every fifth value displayed, from:

```
n=21;rl=0;ru=1;
r=seq(from=rl,to=ru,by=(ru-rl)/(n-1));
```

of Listing 4.1.

- The values $t = 0, 2.0 \times 10^7, 4.0 \times 10^7$ correspond to:

```
dt=2.0e+06;np=21;ip=1;
```

from Listing 4.1 for ip=1,11,21.

- The outer radius remains at $r_u = 1$ throughout the solution as expected for ncase=1, as also reflected in Figure 4.1-7 (for the 21 values of t).

- The derivative drdt=0 corresponding to ncase=1 is confirmed in Figure 4.1-8 (for the 21 values of t).

- The computational effort is appreciable, ncall=3851, which is probably due in part to the multiple starts of ode (np=21, Listing 4.1).

The 2D plots from matplot are in Figures 4.1-1 through 4.1-3.

Figure 4.1-1 indicates that $u_1(r, t)$ initially is a Gaussian and approaches 0.4 as a steady state (equilibrium solution). That is, the cancer cells are spatially uniform as a consequence of the linear, chemotaxis, and haptotaxis diffusion.

Figure 4.1-2 indicates $u_2(r, t)$ has a spatially uniform value of approximately 0.23 at $t = 4.0 \times 10^7$, but this value appears to still be changing with increasing t.

Figure 4.1-3 indicates $u_3(r, t)$ has a significant variation in r at $t = 4.0 \times 10^7$.

Figures 4.1-1 through 4.1-3 generally reflect the complicated form of the solution of eqs. (4.3) and (4.4).

The 3D plots from persp in Figures 4.1-4 through 4.1-6 correspond to the 2D solutions of Figures 4.1-1 through 4.1-3.

Figure 4.1-4 indicates that the concentration of cancer cells goes through a minimum. That is, the cells resurge with increasing t near $r = r_l = 0$.

Figure 4.1-7 indicates no variation in r_u.

Figure 4.1-8 indicates drdt=0.

In summary, the solution has the expected values $r = r_u = 1$, $drdt = 0$ in Figures 4.1-7 and 4.1-8.

ncase=2 is considered next (just the change in ncase from 1 to 2 in Listing 4.1). Abbreviated output follows (Table 4.5).

We can note the following details about this output (including the graphical output in Figures 4.2-7 and 4.2-8).

- The ICs for $u_1(r, t = 0)$, $u_2(r, t = 0)$, $u_3(r, t = 0)$ are the same as for ncase=1.

- Initially ($t = 0$), the values $r = 0, 0.25, 0.5, 0.75, 1$ correspond to 21 values of r over $r_l = 0 \le r \le r_u = 1$ with every fifth value displayed, from:

```
n=21;rl=0;ru=1;
r=seq(from=rl,to=ru,by=(ru-rl)/(n-1));
```

of Listing 4.1.

TABLE 4.5

Abbreviated output from the routines in
Listings 4.1 and 4.2, `ncase=2`, `npara=2`

t	r	u1(r,t)
t	r	u2(r,t)
t	r	u3(r,t)
0.00e+00	0.00	1.0000
0.00e+00	0.00	0.0000
0.00e+00	0.00	0.0000
0.00e+00	0.25	0.5353
0.00e+00	0.25	0.0000
0.00e+00	0.25	0.4647
0.00e+00	0.50	0.0821
0.00e+00	0.50	0.0000
0.00e+00	0.50	0.9179
0.00e+00	0.75	0.0036
0.00e+00	0.75	0.0000
0.00e+00	0.75	0.9964
0.00e+00	1.00	0.0000
0.00e+00	1.00	0.0000
0.00e+00	1.00	1.0000
t	r	u1(r,t)
t	r	u2(r,t)
t	r	u3(r,t)
2.00e+07	0.00	0.1476
2.00e+07	0.00	0.0907
2.00e+07	0.00	0.0000
2.00e+07	0.30	0.1195
2.00e+07	0.30	0.0776
2.00e+07	0.30	0.0092
2.00e+07	0.60	0.0761
2.00e+07	0.60	0.0532
2.00e+07	0.60	0.1251
2.00e+07	0.90	0.0591
2.00e+07	0.90	0.0363

(Continued)

TABLE 4.5 (*Continued*)
Abbreviated output from the routines in
Listings 4.1 and 4.2, ncase=2, npara=2

2.00e+07	0.90	0.4324
2.00e+07	1.20	0.0534
2.00e+07	1.20	0.0310
2.00e+07	1.20	0.5166
t	r	u1(r,t)
t	r	u2(r,t)
t	r	u3(r,t)
4.00e+07	0.00	0.3377
4.00e+07	0.00	0.1922
4.00e+07	0.00	0.0000
4.00e+07	0.35	0.3165
4.00e+07	0.35	0.1812
4.00e+07	0.35	0.0000
4.00e+07	0.70	0.2873
4.00e+07	0.70	0.1649
4.00e+07	0.70	0.0013
4.00e+07	1.05	0.2829
4.00e+07	1.05	0.1601
4.00e+07	1.05	0.0049
4.00e+07	1.40	0.2842
4.00e+07	1.40	0.1601
4.00e+07	1.40	0.0038

ncall = 3075

- The values of r increase with t, and at $t = 4.0 \times 10^7$, $r_u = 1.40$ which is expected since for ncase=2, drdt=1.0e-08, so that drdt*t=1.0e-08* 4.0e+07=0.40 (the change in r is 0.40).
- The values $t = 0, 2.0 \times 10^7, 4.0 \times 10^7$ correspond to:

  dt=2.0e+06;np=21;ip=1;

  from Listing 4.1 (with output at ip=0,11,21).
- A comparison of the solutions for $t = 4.0 \times 10^7$ for ncase=1 and ncase=2 indicates the solutions have changed substantially at $r = r_u$.

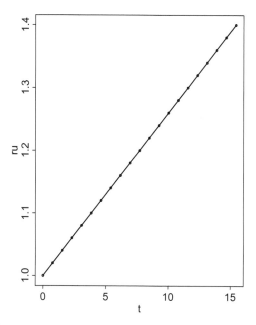

FIGURE 4.2-7
ru from `ru=ru+drdt*dt`, ncase=2.

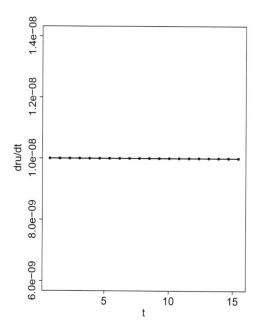

FIGURE 4.2-8
drdt, ncase=2.

```
ncase=1, Table 4.4

4.00e+07   1.00      0.3893
4.00e+07   1.00      0.2293
4.00e+07   1.00      0.0003

ncase=2, Table 4.5

4.00e+07   1.40      0.2842
4.00e+07   1.40      0.1601
4.00e+07   1.40      0.0038
```

However, the graphical solutions for ncase=2 are qualitatively the same as Figures 4.1-1 through 4.1-6 and therefore are not presented here to conserve space.

- The derivative drdt=1.0e-08 corresponding to ncase=2 is confirmed in Figure 4.2-8 (for the 21 values of t).

- The computational effort is ncall=3075.

In summary, Figure 4.2-7 indicates linear variation in r_u with t. Figure 4.2-8 indicates drdt=1.0e-08.

The numerical output for ncase=3 (programmed in Listing 4.1) is considered next. The abbreviated numerical output follows.

TABLE 4.6

Abbreviated output from the routines in
Listings 4.1 and 4.2, ncase=3, npara=2

t	r	u1(r,t)
t	r	u2(r,t)
t	r	u3(r,t)
0.00e+00	0.00	1.0000
0.00e+00	0.00	0.0000
0.00e+00	0.00	0.0000
0.00e+00	0.25	0.5353
0.00e+00	0.25	0.0000
0.00e+00	0.25	0.4647
0.00e+00	0.50	0.0821
0.00e+00	0.50	0.0000
0.00e+00	0.50	0.9179

(Continued)

TABLE 4.6 (*Continued*)

Abbreviated output from the routines in
Listings 4.1 and 4.2, ncase=3, npara=2

0.00e+00	0.75	0.0036
0.00e+00	0.75	0.0000
0.00e+00	0.75	0.9964
0.00e+00	1.00	0.0000
0.00e+00	1.00	0.0000
0.00e+00	1.00	1.0000
t	r	u1(r,t)
t	r	u2(r,t)
t	r	u3(r,t)
2.00e+07	0.00	0.1488
2.00e+07	0.00	0.0915
2.00e+07	0.00	0.0000
2.00e+07	0.34	0.1166
2.00e+07	0.34	0.0766
2.00e+07	0.34	0.0132
2.00e+07	0.67	0.0816
2.00e+07	0.67	0.0543
2.00e+07	0.67	0.1748
2.00e+07	1.01	0.0825
2.00e+07	1.01	0.0454
2.00e+07	1.01	0.3892
2.00e+07	1.34	0.0832
2.00e+07	1.34	0.0444
2.00e+07	1.34	0.2252
t	r	u1(r,t)
t	r	u2(r,t)
t	r	u3(r,t)
4.00e+07	0.00	0.3912
4.00e+07	0.00	0.2295
4.00e+07	0.00	0.0000

(*Continued*)

TABLE 4.6 (*Continued*)
Abbreviated output from the routines in
Listings 4.1 and 4.2, ncase=3, npara=2

4.00e+07	0.38	0.3836
4.00e+07	0.38	0.2259
4.00e+07	0.38	0.0000
4.00e+07	0.76	0.3861
4.00e+07	0.76	0.2274
4.00e+07	0.76	0.0003
4.00e+07	1.13	0.4047
4.00e+07	1.13	0.2373
4.00e+07	1.13	0.0003
4.00e+07	1.51	0.4137
4.00e+07	1.51	0.2422
4.00e+07	1.51	0.0000

ncall = 2976

We can note the following details about this output (including the graphical output in Figures 4.3-4 through 4.3-8).

- The ICs for $u_1(r, t = 0)$, $u_2(r, t = 0)$, $u_3(r, t = 0)$ are the same as for ncase=1,2.
- Initially ($t = 0$), the values $r = 0, 0.25, 0.5, 0.75, 1$ correspond to 21 values of r over $r_l = 0 \leq r \leq r_u = 1$ with every fifth value displayed, from:

  ```
  n=21;rl=0;ru=1;
  r=seq(from=rl,to=ru,by=(ru-rl)/(n-1));
  ```

  of Listing 4.1.
- The values of r increase with t, and at $t = 4.0 \times 10^7$, $r_u = 1.51$ which for ncase=3 follows from drdt=kru*u1[n]*(1-u1[n]-u3[n])*u3[n].
- The values $t = 0, 2.0 \times 10^7, 4.0 \times 10^7$ correspond to:

  ```
  dt=2.0e+06;np=21;ip=1;
  ```

  from Listing 4.1.

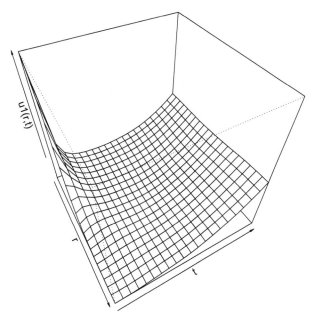

FIGURE 4.3-4
$u_1(r,t)$, ncase=3, npara=2.

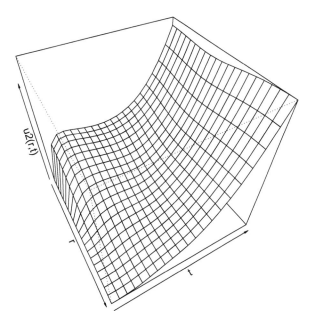

FIGURE 4.3-5
$u_2(r,t)$, ncase=3, npara=2.

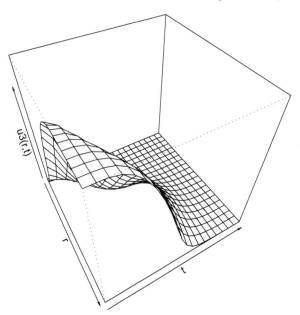

FIGURE 4.3-6
$u_3(r,t)$, ncase=3, npara=2.

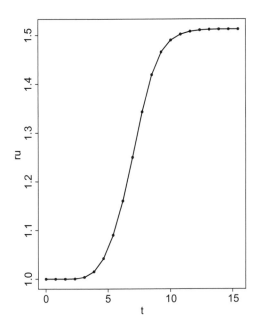

FIGURE 4.3-7
ru from ru=ru+drdt*dt, ncase=3, npara=2.

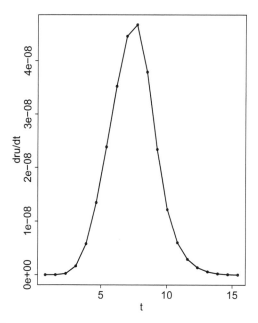

FIGURE 4.3-8
drdt, ncase=3, npara=2.

- A comparison of the solutions for $t = 4.0 \times 10^7$ for ncase=1,2,3 indicates a substantial change.

  ncase=1, Table 4.4

  ```
  4.00e+07   1.00      0.3893
  4.00e+07   1.00      0.2293
  4.00e+07   1.00      0.0003
  ```

  ncase=2, Table 4.5

  ```
  4.00e+07   1.40      0.2842
  4.00e+07   1.40      0.1601
  4.00e+07   1.40      0.0038
  ```

  ncase=3, Table 4.6

  ```
  4.00e+07   1.51      0.4137
  4.00e+07   1.51      0.2422
  4.00e+07   1.51      0.0000
  ```

- The computational effort is reflected in ncall=2976.

The 2D plots of $u_1(r,t)$, $u_2(r,t)$, and $u_3(r,t)$ are not given here. The 3D plots are in Figures 4.3-4 to 4.3-6 and the movement of $r = r_u$ are is given in Figures 4.3-7 and 4.3-8.

Figure 4.3-4 indicates that the cancer cell density near $r = r_l = 0$ goes through a minimum, with a resurgence for later t.

Figure 4.3-7 indicates nonlinear variation in r_u with t. The increase in the tumor radius is 51% (Table 4.6) or a 244% increase in tumor volume.

Figure 4.3-8 indicates the t variation in drdt.

Overall, Figures 4.3-4 through 4.3-8 indicate a complicated solution of eqs. (4.3) through (4.6). Further insight into the source of this complexity is available by computing and displaying the RHS and LHS terms of eqs. (4.3) and (4.4) as explained in Chapter 3 of [6]. This is left as an exercise.

4.5 Summary and Conclusions

The tumor model of eqs. (4.3) through (4.6) is implemented in the two R routines of Listings 4.1 and 4.2 within the MOL framework. The three cases ncase=1,2,3 demonstrate the effect of the tumor outer boundary (r_u) movement. For example, for the parameters specified with npara=2 in Listing 4.1 with ncase=3, the tumor boundary moves 51% with the boundary velocity function drdt=kru*u1[n]*(1-u1[n]-u3[n])*u3[n]. Alternate equations of boundary motion can easily be investigated with changes in the main program of Listing 4.1.

As a concluding example application of a moving boundary PDE model, the formation of plaque from atherosclerosis is considered in the next chapter.

References

1. Bird, R.B., W.E. Stewart, and E.N. Lightfoot (2002), *Transport Phenomena*, 2nd ed., John Wiley & Sons, Hoboken, NJ, p. 836.

2. Pang, P.Y.H., and Y. Wang (2018), Global boundedness of solutions to a chemotaxis-haptotaxis model with tissue remodeling, *Mathematical Modeling and Methods in Applied Science*, **28**, no. 11, 2211–2235.

3. Soetaert, K., J. Cash, and F. Mazzia (2012), *Solving Differential Equations in R*, Springer-Verlag, Heidelberg, Germany, pp. 140–141.

4. Schiesser, W.E. (2013), *Partial Differential Equation Analysis in Biomedical Engineering*, Cambridge University Press, Cambridge, UK.

5. Schiesser, W.E. (2014), *Differential Equation Analysis in Biomedical Science and Engineering: Partial Differential Equation Applications in R*, John Wiley & Sons, Hoboken, NJ.

6. Schiesser, W.E. (2018), *A Mathematical Modeling Approach to Infectious Diseases: Cross Diffusion PDE Models for Epidemiology*, World Scientific Publishing Company, Singapore.

5

Plaque Formation in Atherosclerosis

5.1   Introduction

The moving boundary partial differential equation (MBPDE) application that follows pertains to arterial plaque formation in atherosclerosis. Specifically, as plaque forms on the epithelium inner layer (EIL, intima), the inner arterial wall moves toward the center of the artery resulting in reduced blood flow (as well as stiffening of the arterial wall, also termed *hardening of the artery*).

5.2   PDE Model

The mathematical model consists of six 1D partial differential equations (PDEs) as documented in Chapter 1 of [5]. The EIL inner boundary moves according to an equation of motion that is based on the formation of foam cells.

The dependent variables of the six PDEs are listed in Table 5.1. The numerical solution of the PDEs gives the dependent variables as a function of x, the distance across the EIL, and time t, that is, the spatiotemporal variation of the concentrations and densities of the six dependent variables:

A principal output of the model is the spatiotemporal evolution of foam cells as a response to low and high density lipoproteins (LDL, HDL) in the bloodstream. Foam cells in turn lead to arterial stiffness and plaque that contribute to atherosclerosis pathology.

The model is intended as a prototype that the reader/analyst/researcher can study by using the R routines provided with the book. The programmed base case can first be executed to confirm the numerical and graphical output discussed in the book. The model can then be modified through variation of the parameters and extended through the modification of and addition to the PDEs. For example, the transport coefficients defined numerically in the main program of Listing 5.1 (discussed subsequently) can be varied as a function of x to reflect the changing properties as the EIL moving boundary evolves with t.

TABLE 5.1
Dependent variables of the PDE model, [1,5]

$\ell(x,t)$	concentration of modified LDL
$h(x,t)$	concentration of HDL
$p(x,t)$	concentration of chemoattractants
$q(x,t)$	concentration of ES cytokines
$m(x,t)$	density of monocytes/macrophages
$N(x,t)$	density of foam cells

The numerical values of the parameters were selected to give a time scale of two months, and observed diffusion rates across the EIL (since initially the endothelial layer is thin, e.g., $40\mu m$,[1] the variation in x is small).

The six PDEs are generally based on conservation principles, with spatial transport according to Fick's first law for diffusion. For example, for the modified LDL concentration,[2] $\ell(x,t)$ a mass balance on an incremental volume [5] gives:

$$\frac{\partial \ell}{\partial t} = D_\ell \frac{\partial^2 \ell}{\partial x^2} - r_{\ell 1} \frac{\ell}{r_{l3} + \ell} m - r_{\ell 2}\ell. \qquad (5.1\text{-}1)$$

The diffusion flux of modified LDL is given by Fick's first law.

$$q_x = -D_\ell \frac{\partial \ell}{\partial x}$$

The minus is included so that the flux is in the direction of decreasing ℓ (the gradient $\dfrac{\partial \ell}{\partial x}$ is negative).

Eq. (5.1-1) includes a nonlinear rate term which is a function of ℓ and m.

$$-r_{\ell 1} \frac{\ell}{r_{l3} + \ell} m$$

The use of nonlinear functions is straightforward numerically as demonstrated in the subsequent coding.

Eq. (5.1-1) is the first PDE in the system of six PDEs. The terms in Eq. (5.1-1) apply to the interior of the EIL adjacent to the bloodstream [1].

Similarly, a mass balance on the HDL with concentration $h(x,t)$ gives:

$$\frac{\partial h}{\partial t} = D_h \frac{\partial^2 h}{\partial x^2} - r_{h1} \frac{h}{r_{h3} + h} N - r_{h2}h. \qquad (5.1\text{-}2)$$

---

[1] $1\ micron\ (\mu m) = 1\ micrometer = 10^{-6}m = 10^{-4}cm = 10^3\ nanometers\ (nm)$. The model is expressed in cgs units.

[2] The LDL in the blood stream when transported to the EIL is modified, for example, by oxidation. It is therefore termed *modified LDL*, and its concentration is designated as ℓ.

The terms in Eq. (5.1-2) again apply to the EIL adjacent to the blood-stream [1].

A mass balance for the chemoattractants with concentration $p(x,t)$, ES cytokines with concentration $q(x,t)$, macrophages with density $m(x,t)$, and foam cells with density $N(x,t)$ gives

$$\frac{\partial p}{\partial t} = D_p \frac{\partial^2 p}{\partial x^2} + r_{p1} \frac{\ell}{r_{l3} + \ell} m - r_{p2} p \tag{5.1-3}$$

$$\frac{\partial q}{\partial t} = D_q \frac{\partial^2 q}{\partial x^2} + r_{q1} \frac{\ell}{r_{l3} + \ell} m - r_{q2} q. \tag{5.1-4}$$

The mass balance for the macrophages with density $m(x,t)$ includes chemo-taxis diffusion. The net diffusion flux is:

$$-r_{m1} \frac{\partial}{\partial x} (m \frac{\partial \ell}{\partial x}),$$

which expands to

$$-r_{m1} \left(\frac{\partial m}{\partial x} \frac{\partial \ell}{\partial x} + m \frac{\partial^2 \ell}{\partial x^2} \right).$$

The diffusivity r_{m1} is a constant, but it multiplies $m(x,t)$ so that this flux is nonlinear, with the derivative (gradient) $\frac{\partial \ell}{\partial x}$.

This expanded form of the chemotaxis diffusion term is then used in the mass balance for $m(x,t)$[3]:

$$\frac{\partial m}{\partial t} = D_m \frac{\partial^2 m}{\partial x^2} - r_{m1}(\frac{\partial m}{\partial x} \frac{\partial l}{\partial x} + m \frac{\partial^2 l}{\partial x^2})$$

$$-r_{m2} \frac{\ell}{r_{l3} + \ell} m + r_{m3} \frac{h}{r_{h3} + h} N - r_{m4} m + r_{m5} \ell. \tag{5.1-5}$$

Finally, for the foam cell density $N(x,t)$:

$$\frac{\partial N}{\partial t} = D_N \frac{\partial^2 N}{\partial x^2} + r_{N1} \frac{\ell}{r_{l3} + \ell} m - r_{N2} \frac{h}{r_{h3} + h} N + r_{N3} \ell. \tag{5.1-6}$$

---

[3]The terms $r_{m5}\ell$ and $r_{N3}\ell$ were added by the author (WES) to Eqs. (5.1-5) and (5.1-6), respectively, to give nonzero RHS terms. Otherwise, if all of the RHS terms in Eqs. (5.1-5) and (5.1-6) are zero initially (according to ICs (5.3) discussed subsequently), they remain at zero so the solutions $m(x,t), N(x,t)$, as well as $p(x,t), q(x.t)$, do not move away from the homogeneous (zero) ICs. This special case is demonstrated subsequently.

Eqs. (5.1) constitute the set of six PDEs. Each of the PDEs (5.1) is first order in t and second order in x. Therefore, each requires one initial condition (IC) and two boundary conditions (BCs). The BCs follow as Eqs. (5.2-1) through (5.2-12)[4]:

$$D_\ell \frac{\partial \ell(x = x_l, t)}{\partial x} = -k_l(\ell_0 - \ell(x = x_l, t)); \quad \frac{\partial \ell(x = x_u, t)}{\partial x} = 0 \quad (5.2\text{-}1, 5.2\text{-}2)$$

$$D_h \frac{\partial h(x = x_l, t)}{\partial x} = -k_h(h_0 - h(x = x_l, t)); \quad \frac{\partial h(x = x_u, t)}{\partial x} = 0 \quad (5.2\text{-}3, 5.2\text{-}4)$$

$$D_p \frac{\partial p(x = x_l, t)}{\partial x} = -k_p(p_0 - p(x = x_l, t)); \quad \frac{\partial p(x = x_u, t)}{\partial x} = 0 \quad (5.2\text{-}5, 5.2\text{-}6)$$

$$D_q \frac{\partial q(x = x_l, t)}{\partial x} = -k_q(q_0 - q(x = x_l, t)); \quad \frac{\partial q(x = x_u, t)}{\partial x} = 0 \quad (5.2\text{-}7, 5.2\text{-}8)$$

$$D_m \frac{\partial m(x = x_l, t)}{\partial x} = -k_m(m_0 - m(x = x_l, t)); \quad \frac{\partial m(x = x_u, t)}{\partial x} = 0 \\ (5.2\text{-}9, 5.2\text{-}10)$$

$$D_N \frac{\partial N(x = x_l, t)}{\partial x} = -k_N(N_0 - N(x = x_l, t)); \quad \frac{\partial N(x = x_u, t)}{\partial x} = 0. \\ (5.2\text{-}11, 5.2\text{-}12)$$

The inner boundary of the EIL, $x = x_l$, moves according to the equation of motion:

$$\frac{dx_l}{dt} = -k_{x_l} N(x = x_l, t). \quad (5.2\text{-}13)$$

The velocity is proportional to the foam cell density at the inner boundary of the EIL. The velocity is negative, that is, the EIL thickness increases toward the center of the artery. The starting thickness is 40 μm and increases as atherosclerosis progresses.

---

[4]Conventional Robin BCs based on continuity of mass transfer rates are specified first for each PDE at $x = x_l$. The mass transfer rates across this surface can be adjusted through the selection of coefficients k_l to k_N as demonstrated subsequently.

The BCs at $x = x_u$ are homogeneous (zero flux) Neumann BCs.

The BCs stated in [1, 2] can be used based on the boundary concentrations $\ell(x = x_l, t)$ to $N(x = x_l, t)$. Attempts to do this produced unrealistic/unacceptable results, e.g., negative and/or large concentrations. If the reader/analyst/researcher implements these surface BCs, additional parameters will have to be specified.

The ICs for Eqs. (5.1) follow:

$$\ell(x, t = 0) = f_\ell(x) \tag{5.3-1}$$

$$h(x, t = 0) = f_h(x) \tag{5.3-2}$$

$$p(x, t = 0) = f_p(x) \tag{5.3-3}$$

$$q(x, t = 0) = f_q(x) \tag{5.3-4}$$

$$m(x, t = 0) = f_m(x) \tag{5.3-5}$$

$$N(x, t = 0) = f_N(x). \tag{5.3-6}$$

f_h to f_N are functions to be specified. For the subsequent discussion, they are taken as the zero function.

The method of lines (MOL) implementation of Eqs. (5.1) through (5.3) in a series of R routines is considered next.

5.2.1 Main Program

The following main program is an extension of the main program in Listing 2.1 for six PDEs rather than one.

Listing 5.1: Main program for Eqs. (5.1) through (5.3)

```
#
# Six PDE atherosclerosis model
#
# Delete previous workspaces
  rm(list=ls(all=TRUE))
#
# Access ODE integrator
  library("deSolve");
#
# Access functions for numerical solution
  setwd("f:/mbpde/chap5");
  source("pde_1a.R");
#
# Select case for moving boundary
  ncase=1;
```

(Continued)

Listing 5.1 (Continued): Main program for Eqs. (5.1) through (5.3)

```
#
# Parameters
#
# Diffusivities
  D_l=1.0e-08; D_h=1.0e-08;
  D_p=1.0e-08; D_q=1.0e-08;
  D_m=1.0e-08; D_N=1.0e-08;
#
# Mass transfer coefficients
  k_l=1.0e-08; k_h=1.0e-08;
  k_p=0.0e-08; k_q=1.0e-08;
  k_m=0.0e-08; k_N=0.0e-08;
#
# Bloodstream (lumen) concentrations
  l0=1; h0=3;
  p0=0; q0=0;
  m0=0; N0=0;
#
# Reaction kinetic rate constants
#
# l(x,t)
  r_l1=5.0e-08; r_l2=5.0e-08;
  r_l3=1.0e-05;
#
# h(x,t);
  r_h1=5.0e-08; r_h2=5.0e-08;
  r_h3=1.0e-05;
#
# p(x,t)
  r_p1=5.0e-08; r_p2=5.0e-08;
#
# q(x,t)
  r_q1=5.0e-08; r_q2=5.0e-08;
#
# m(x,t)
  r_m1=5.0e-08; r_m2=5.0e-08;
  r_m3=5.0e-08; r_m4=5.0e-08;
  r_m5=5.0e-08;
```

(Continued)

Listing 5.1 (Continued): Main program for Eqs. (5.1) through (5.3)

```
#
# N(x,t)
  r_N1=5.0e-08; r_N2=5.0e-08;
  r_N3=5.0e-08;
#
# kx1
  kx1=1.0e-08;
#
# Grid (in x)
  nx=26;
  xl=0;xu=0.004;
  x=seq(from=xl,to=xu,(xu-xl)/(nx-1));
#
# Independent variable for ODE integration
  t0=0;
  tout=rep(0,2);
  dt=1.0e+06;np=11;ip=1;
  tp=rep(0,np);xlp=rep(0,np);
  dxdtp=rep(0,(np-1));
#
# IC for first step in t
  u=rep(0,nrow=6*nx);
  u0=rep(0,6*nx);
  l=rep(0,nx);h=rep(0,nx);
  p=rep(0,nx);q=rep(0,nx);
  m=rep(0,nx);N=rep(0,nx);
 lp=matrix(0,nrow=nx,ncol=np);
 hp=matrix(0,nrow=nx,ncol=np);
 pp=matrix(0,nrow=nx,ncol=np);
 qp=matrix(0,nrow=nx,ncol=np);
 mp=matrix(0,nrow=nx,ncol=np);
 Np=matrix(0,nrow=nx,ncol=np);
  u0=rep(0,6*nx);
  for(i in 1:nx){
        u0[i]=0;l[i]=u0[i];        lp[i,1]=l[i]
      u0[i+nx]=0;h[i]=u0[i+nx];   hp[i,1]=h[i]
    u0[i+2*nx]=0;p[i]=u0[i+2*nx];pp[i,1]=p[i]
    u0[i+3*nx]=0;q[i]=u0[i+3*nx];qp[i,1]=q[i]
    u0[i+4*nx]=0;m[i]=u0[i+4*nx];mp[i,1]=m[i]
    u0[i+5*nx]=0;N[i]=u0[i+5*nx];Np[i,1]=N[i]
  }
```

(Continued)

**Listing 5.1 (Continued): Main program for Eqs. (5.1)
through (5.3)**

```
  tp[1]=t0;xlp[1]=xl;
  ncall=0;
#
# Display IC
  cat(sprintf("\n\n                t             x          l(x,t)"));
  cat(sprintf("\n              t            x         h(x,t)"));
  cat(sprintf("\n              t            x         p(x,t)"));
  cat(sprintf("\n              t            x         q(x,t)"));
  cat(sprintf("\n              t            x         m(x,t)"));
  cat(sprintf("\n              t            x         N(x,t)"));
   iv=seq(from=1,to=nx,by=5);
   for(i in iv){
     cat(sprintf("\n %12.3e%12.3e%12.3e",t0,x[i],l[i]));
     cat(sprintf("\n %12.3e%12.3e%12.3e",t0,x[i],h[i]));
     cat(sprintf("\n %12.3e%12.3e%12.3e",t0,x[i],p[i]));
     cat(sprintf("\n %12.3e%12.3e%12.3e",t0,x[i],q[i]));
     cat(sprintf("\n %12.3e%12.3e%12.3e",t0,x[i],m[i]));
     cat(sprintf("\n %12.3e%12.3e%12.3e\n",t0,x[i],N[i]));
   }
#
# Next step along solution
  while(ip<np){
  for(i in 1:nx){
        u0[i]=l[i];
      u0[i+nx]=h[i];
    u0[i+2*nx]=p[i];
    u0[i+3*nx]=q[i];
    u0[i+4*nx]=m[i];
    u0[i+5*nx]=N[i];
  }
  t0=tout[2];
  tout[1]=t0;
  tout[2]=tout[1]+dt;
#
# ODE integration
  out=ode(func=pde_1a,y=u0,times=tout);
#
# Arrays for solution
  for(i in 1:nx){
    l[i]=out[2,i+1];
```

(Continued)

Listing 5.1 (Continued): Main program for Eqs. (5.1) through (5.3)

```
    h[i]=out[2,i+1+nx];
    p[i]=out[2,i+1+2*nx];
    q[i]=out[2,i+1+3*nx];
    m[i]=out[2,i+1+4*nx];
    N[i]=out[2,i+1+5*nx];
  }
#
# Redefine spatial grid
  tablel=splinefun(x,l);
  tableh=splinefun(x,h);
  tablep=splinefun(x,p);
  tableq=splinefun(x,q);
  tablem=splinefun(x,m);
  tableN=splinefun(x,N);
  if(ncase==1){dxdt=0;}
  if(ncase==2){dxdt=-1.0e-10;}
  if(ncase==3){dxdt=-kxl*N[1];}
  xl=xl+dxdt*dt;
  x=seq(from=xl,to=xu,by=(xu-xl)/(nx-1));
#
# Solution on redefined grid
  l=tablel(x,deriv=0);
  h=tableh(x,deriv=0);
  p=tablep(x,deriv=0);
  q=tableq(x,deriv=0);
  m=tablem(x,deriv=0);
  N=tableN(x,deriv=0);
  ip=ip+1;
  xlp[ip]=xl;dxdtp[ip-1]=dxdt;
#
# Display numerical solution
# cat(sprintf("\n ip = %2d\n",ip));
  cat(sprintf("\n\n            t          x       l(x,t)"));
  cat(sprintf("\n         t        x       h(x,t)"));
  cat(sprintf("\n         t        x       p(x,t)"));
  cat(sprintf("\n         t        x       q(x,t)"));
  cat(sprintf("\n         t        x       m(x,t)"));
  cat(sprintf("\n         t        x       N(x,t)"));
  iv=seq(from=1,to=nx,by=5);
```

(Continued)

Listing 5.1 (Continued): Main program for Eqs. (5.1) through (5.3)

```
  for(i in iv){
    cat(sprintf("\n %12.3e%12.3e%12.3e",tout[2],x[i],l[i]));
    cat(sprintf("\n %12.3e%12.3e%12.3e",tout[2],x[i],h[i]));
    cat(sprintf("\n %12.3e%12.3e%12.3e",tout[2],x[i],p[i]));
    cat(sprintf("\n %12.3e%12.3e%12.3e",tout[2],x[i],q[i]));
    cat(sprintf("\n %12.3e%12.3e%12.3e",tout[2],x[i],m[i]));
    cat(sprintf("\n %12.3e%12.3e%12.3e\n",tout[2],x[i],N[i]));
  }
#
# Solution for plotting
  for(i in 1:nx){
    lp[i,ip]=l[i];
    hp[i,ip]=h[i];
    pp[i,ip]=p[i];
    qp[i,ip]=q[i];
    mp[i,ip]=m[i];
    Np[i,ip]=N[i];
  }
    tp[ip]=tout[2];
#
# Next step (from while)
  }
#
# Plot output
#
# 3D
  persp(x,tp,lp,theta=60,phi=45,
        xlim=c(xlp[np],xu),ylim=c(tp[1],tp[np]),xlab="x",
        ylab="t",zlab="l(x,t)");
  persp(x,tp,hp,theta=60,phi=45,
        xlim=c(xlp[np],xu),ylim=c(tp[1],tp[np]),xlab="x",
        ylab="t",zlab="h(x,t)");
  persp(x,tp,pp,theta=60,phi=45,
        xlim=c(xlp[np],xu),ylim=c(tp[1],tp[np]),xlab="x",
        ylab="t",zlab="p(x,t)");
  persp(x,tp,qp,theta=60,phi=45,
        xlim=c(xlp[np],xu),ylim=c(tp[1],tp[np]),xlab="x",
        ylab="t",zlab="q(x,t)");
  persp(x,tp,mp,theta=60,phi=45,
        xlim=c(xlp[np],xu),ylim=c(tp[1],tp[np]),xlab="x",
        ylab="t",zlab="m(x,t)");
```

(Continued)

Listing 5.1 (Continued): Main program for Eqs. (5.1)
through (5.3)

```
  persp(x,tp,Np,theta=60,phi=45,
       xlim=c(xlp[np],xu),ylim=c(tp[1],tp[np]),xlab="x",
       ylab="t",zlab="N(x,t)");
#
# Boundary position
  mon=60*60*24*30;
  mon=1;
  plot(tp/mon,xlp,xlab="t",ylab="xl");
    lines(tp/mon,xlp,type="l",lwd=2);
  points(tp/mon,xlp,pch="o",lwd=2);
  plot(tp[2:np]/mon,dxdtp,xlab="t",ylab="dru/dt");
    lines(tp[2:np]/mon,dxdtp,type="l",lwd=2);
  points(tp[2:np]/mon,dxdtp,pch="o",lwd=2);
#
# Calls to ODE routine
  cat(sprintf("\n\n  ncall = %3d\n",ncall));
```

We can note the following details about Listing 5.1.

- Previous workspaces are deleted.

```
  #
  # Six PDE atherosclerosis model
  #
  # Delete previous workspaces
    rm(list=ls(all=TRUE))
```

- The R ODE integrator library deSolve is accessed. Then the directory with the files for the solution of Eqs. (5.1) through (5.3) is designated. Note that setwd (set working directory) uses / rather than the
  usual \.

```
  #
  # Access ODE integrator
    library("deSolve");
  #
  # Access functions for numerical solution
    setwd("f:/mbpde/chap5");
    source("pde_1a.R");
```

  pde_1a.R is the routine for the MOL approximation of PDEs (5.1)
  (discussed subsequently).

- ncase is specified with three possible values, ncase=1,2,3, corresponding to different velocities for the moving boundary at $r = x_l$).

  ```
  # Select case for moving boundary
     ncase=1;
  ```

- The model parameters are defined.

  ```
  #
  # Parameters
  #
  # Diffusivities
     D_l=1.0e-08; D_h=1.0e-08;
     D_p=1.0e-08; D_q=1.0e-08;
     D_m=1.0e-08; D_N=1.0e-08;
  #
  # Mass transfer coefficients
     k_l=1.0e-08; k_h=1.0e-08;
     k_p=0.0e-08; k_q=1.0e-08;
     k_m=0.0e-08; k_N=0.0e-08;
  #
  # Bloodstream (lumen) concentrations
     l0=1; h0=3;
     p0=0; q0=0;
     m0=0; N0=0;
  #
  # Reaction kinetic rate constants
  #
  # l(x,t)
     r_l1=5.0e-08; r_l2=5.0e-08;
     r_l3=1.0e-05;
  #
  # h(x,t);
     r_h1=5.0e-08; r_h2=5.0e-08;
     r_h3=1.0e-05;
  #
  # p(x,t)
     r_p1=5.0e-08; r_p2=5.0e-08;
  #
  # q(x,t)
     r_q1=5.0e-08; r_q2=5.0e-08;
  #
  # m(x,t)
     r_m1=5.0e-08; r_m2=5.0e-08;
     r_m3=5.0e-08; r_m4=5.0e-08;
     r_m5=5.0e-08;
  ```

```
#
# N(x,t)
  r_N1=5.0e-08; r_N2=5.0e-08;
  r_N3=5.0e-08;
#
# kx1
  kx1=1.0e-08;
```

These parameters are used in Eqs. (5.1) and (5.2). In particular, k_{x_l} = kx1 = 1.0e-08 in Eq. (5.2-13).

- A spatial grid of 26 points is defined for $r_l = 0 \leq r \leq x_l = 1$ so that $r = 0, 0.004/25, ..., 0.004$.

```
#
# Grid (in x)
  nx=26;
  xl=0;xu=0.004;
  x=seq(from=xl,to=xu,(xu-xl)/(nx-1));
```

x_l = xl, the lower limit on x, is changed, reflecting the moving inner boundary for the MBPDEs.

- Parameters for the MOL solution are defined.

```
#
# Independent variable for ODE integration
  t0=0;
  tout=rep(0,2);
  dt=1.0e+06;np=11;ip=1;
  tp=rep(0,np);xlp=rep(0,np);
  dxdtp=rep(0,(np-1));
```

These statements require some additional explanation.

- The initial value of t for the solution is defined.

    t0=0

- Rather than call **ode** once to compute a complete solution from t_0 to a final time t_f, **ode** is called for a series of output points in vector **tout**. In each of these intervals of two points, the grid in x is defined for an updated x_l (calculated by xl=xl+dxdt*dt). In this way, x_l is refined as the solution proceeds to reflect the moving boundary.

    tout=rep(0,2);

- Each interval of two points has length dt=1.0e+06 and 11 output points are defined, np=11 (including $t = t_0$). Therefore, the total interval in t for the complete solution is $(11 - 1)(1.0e + 06) = 1.0e + 07$. The first point in each interval has index ip=1, and the second point has index ip=2.

```
dt=1.0e+06;np=11;ip=1;
```

- The value of t at the 11 output points is placed in vector tp and the corresponding values of x_l are placed in vector xlp for plotting. In this way, the movement of the boundary at $x = x_l$ can be observed graphically.

```
tp=rep(0,np);xlp=rep(0,np);
```

- Similarly, the varying values of $\dfrac{dx_l}{dt}$ (for ncase=1,2,3) are placed in dxdtp for plotting. Since this derivative is not available initially at $t = t_0$ (but only after xl=xl+dxdt*dt is used), there are $11-1 = 10$ values of the derivative.

```
dxdtp=rep(0,(np-1));
```

- Functions $h_\ell(x)$ to $h_N(x)$ in ICs (5.3) are defined as the zero function. This is the beginning of the coding for the simultaneous PDEs (Eqs. (5.1)).

```
#
# IC for first step in t
  u=rep(0,nrow=6*nx);
  u0=rep(0,6*nx);
  l=rep(0,nx);h=rep(0,nx);
  p=rep(0,nx);q=rep(0,nx);
  m=rep(0,nx);N=rep(0,nx);
 lp=matrix(0,nrow=nx,ncol=np);
 hp=matrix(0,nrow=nx,ncol=np);
 pp=matrix(0,nrow=nx,ncol=np);
 qp=matrix(0,nrow=nx,ncol=np);
 mp=matrix(0,nrow=nx,ncol=np);
 Np=matrix(0,nrow=nx,ncol=np);
  u0=rep(0,6*nx);
  for(i in 1:nx){
        u0[i]=0;l[i]=u0[i];        lp[i,1]=l[i]
       u0[i+nx]=0;h[i]=u0[i+nx];   hp[i,1]=h[i]
      u0[i+2*nx]=0;p[i]=u0[i+2*nx];pp[i,1]=p[i]
      u0[i+3*nx]=0;q[i]=u0[i+3*nx];qp[i,1]=q[i]
      u0[i+4*nx]=0;m[i]=u0[i+4*nx];mp[i,1]=m[i]
      u0[i+5*nx]=0;N[i]=u0[i+5*nx];Np[i,1]=N[i]
  }
  tp[1]=t0;xlp[1]=xl;
  ncall=0;
```

The six IC functions are placed in a single vector u0 for the start of the integration of 6*n = 6*26 = 156 MOL ODEs. Also, the ICs are placed in l, h, p, q, m, N for subsequent use in the moving boundary algorithm, and in lp, hp, pp, hp, mp, Np for plotting. The initial values $t =$ tp[1] and $x_l =$ xlp[1] are also defined. The counter for the calls to the ODE/MOL routine pde_1a is initialized.

- The solutions $\ell(x, t = 0)$ to $N(x, t = 0)$ are displayed for every fifth value of the $n = 26$ values of x (using by=5).

```
#
# Display IC
    cat(sprintf("\n\n                 t           x        l(x,t)"));
    cat(sprintf("\n          t           x        h(x,t)"));
    cat(sprintf("\n          t           x        p(x,t)"));
    cat(sprintf("\n          t           x        q(x,t)"));
    cat(sprintf("\n          t           x        m(x,t)"));
    cat(sprintf("\n          t           x        N(x,t)"));
    iv=seq(from=1,to=nx,by=5);
    for(i in iv){
      cat(sprintf("\n %12.3e%12.3e%12.3e",t0,x[i],l[i]));
      cat(sprintf("\n %12.3e%12.3e%12.3e",t0,x[i],h[i]));
      cat(sprintf("\n %12.3e%12.3e%12.3e",t0,x[i],p[i]));
      cat(sprintf("\n %12.3e%12.3e%12.3e",t0,x[i],q[i]));
      cat(sprintf("\n %12.3e%12.3e%12.3e",t0,x[i],m[i]));
      cat(sprintf("\n %12.3e%12.3e%12.3e\n",t0,x[i],N[i]));
    }
```

- The next interval of two points is initialized. For ip=1, the first interval $0 \le t \le dt = 1.0e + 06$ is initialized (using the while).

```
#
# Next step along solution
    while(ip<np){
    for(i in 1:nx){
          u0[i]=l[i];
        u0[i+nx]=h[i];
      u0[i+2*nx]=p[i];
      u0[i+3*nx]=q[i];
      u0[i+4*nx]=m[i];
      u0[i+5*nx]=N[i];
    }
    t0=tout[2];
    tout[1]=t0;
    tout[2]=tout[1]+dt;
```

- The system of $n = 156$ MOL/ODEs is integrated by the library integrator ODE (available in deSolve [6]). As expected, the inputs to ODE are the ODE function, pde_1a, the IC vector u0, and the vector of output values of t, tout. The length of u0 ($6(26) = 156$) informs ODE how many ODEs are to be integrated. func,y,times are reserved names.

```
#
# ODE integration
  out=ode(func=pde_1a,y=u0,times=tout);
```

The numerical solution to the ODEs is returned in matrix out. In this case, out has the dimensions $2 \times 6nx + 1 = 2 \times 156 + 1 = 157$.

The offset $156+1$ is required since the first element of each column has the output t (also in tout), and the $2, ..., 6nx + 1 = 2, ..., 157$ column elements have the 156 ODE solutions.

- The solution of the 156 ODEs returned in out by ODE is placed in vectors l, h, p, q, m, N of length 26 (defined previously).

```
#
# Arrays for solution
  for(i in 1:nx){
    l[i]=out[2,i+1];
    h[i]=out[2,i+1+nx];
    p[i]=out[2,i+1+2*nx];
    q[i]=out[2,i+1+3*nx];
    m[i]=out[2,i+1+4*nx];
    N[i]=out[2,i+1+5*nx];
  }
```

- xl=xl+dxdt*dt is used to redefine x_l.

```
#
# Redefine spatial grid
  tablel=splinefun(x,l);
  tableh=splinefun(x,h);
  tablep=splinefun(x,p);
  tableq=splinefun(x,q);
  tablem=splinefun(x,m);
  tableN=splinefun(x,N);
  if(ncase==1){dxdt=0;}
  if(ncase==2){dxdt=-1.0e-10;}
  if(ncase==3){dxdt=-kxl*N[1];}
  xl=xl+dxdt*dt;
  x=seq(from=xl,to=xu,by=(xu-xl)/(nx-1));
```

This code requires some additional explanation.

- A table of spline coefficients is defined for each dependent variable at the current x with splinefun.

```
table1=splinefun(x,1);
tableh=splinefun(x,h);
tablep=splinefun(x,p);
tableq=splinefun(x,q);
tablem=splinefun(x,m);
tableN=splinefun(x,N);
```

- Three cases for the $\dfrac{dx_l}{dt}$ = dxdt are defined.

```
if(ncase==1){dxdt=0;}
if(ncase==2){dxdt=-1.0e-10;}
if(ncase==3){dxdt=-kxl*N[1];}
```

For ncase=1, $\dfrac{dx_l}{dt} = 0$ so there is no change in x_l. This case is worth considering since if x_l changes, a programming error is indicated.

For ncase=2, x_l changes at a constant rate 1.0e-08 so that the variation of x_l with t is linear. Again, this case is worth considering since if this response of x_l is not observed, a programming error is indicated.

For ncase=3, dxdt is defined by the function -kxl*N[1] in accordance with Eq. (5.2-13). Other choices for the moving boundary function can be programmed here. A particular choice will determine how the artery inner boundary at $x = x_l$ moves with t.

- The explicit Euler method is used to compute the next x_l.

```
xl=xl+dxdt*dt;
```

dt is presumed small enough that the Euler method gives sufficient accuracy in the calculation of x_l.

- The grid in x is redefined for the new x_l. In other words, the moving grid is implemented at this point.

```
x=seq(from=xl,to=xu,by=(xu-xl)/(nx-1));
```

- Modified solutions l, h, p, q, m, N are computed for the redefined grid x. deriv=0 designates the return of the function (rather than its derivatives in x) interpolated/extrapolated by the spline.

```
#
# Solution on redefined grid
l=table1(x,deriv=0);
```

```
h=tableh(x,deriv=0);
p=tablep(x,deriv=0);
q=tableq(x,deriv=0);
m=tablem(x,deriv=0);
N=tableN(x,deriv=0);
```

This step illustrates the use of an important property of the spline, that is, a different independent variable vector x can be defined and used which permits the implementation of the moving grid.

- The current values of x_l and $\dfrac{dx_l}{dt}$ are updated for subsequent plotting.

```
ip=ip+1;
xlp[ip]=xl;dxdtp[ip-1]=dxdt;
```

dxdtp[ip-1] is used since the vector dxdtp does not include a value of the derivative at $t = t_0 = 0$ (discussed previously).

- The solution (returned by **ode** in **out**) is displayed for every fifth value of x (with **by=5**).

```
#
# Display numerical solution
# cat(sprintf("\n ip = %2d\n",ip));
  cat(sprintf("\n\n                 t         x         l(x,t)"));
  cat(sprintf("\n          t         x         h(x,t)"));
  cat(sprintf("\n          t         x         p(x,t)"));
  cat(sprintf("\n          t         x         q(x,t)"));
  cat(sprintf("\n          t         x         m(x,t)"));
  cat(sprintf("\n          t         x         N(x,t)"));
  iv=seq(from=1,to=nx,by=5);
  for(i in iv){
   cat(sprintf("\n %12.3e%12.3e%12.3e",tout[2],x[i],l[i]));
   cat(sprintf("\n %12.3e%12.3e%12.3e",tout[2],x[i],h[i]));
   cat(sprintf("\n %12.3e%12.3e%12.3e",tout[2],x[i],p[i]));
   cat(sprintf("\n %12.3e%12.3e%12.3e",tout[2],x[i],q[i]));
   cat(sprintf("\n %12.3e%12.3e%12.3e",tout[2],x[i],m[i]));
   cat(sprintf("\n %12.3e%12.3e%12.3e\n",tout[2],x[i],N[i]));
  }
```

- The solutions and t are also placed in arrays for plotting.

```
#
# Solution for plotting
  for(i in 1:nx){
    lp[i,ip]=l[i];
    hp[i,ip]=h[i];
```

```
        pp[i,ip]=p[i];
        qp[i,ip]=q[i];
        mp[i,ip]=m[i];
        Np[i,ip]=N[i];
      }
      tp[ip]=tout[2];
```

- The next interval of length `dt` is programmed, including a call to `ode` (for `ip<np`).

```
#
# Next step (from while)
   }
```

- At the end of the final step in t, the solution of Eqs. (5.1) are plotted in 3D with `persp`.

```
#
# Plot output
#
# 3D
  persp(x,tp,lp,theta=60,phi=45,
        xlim=c(xlp[np],xu),ylim=c(tp[1],tp[np]),xlab="x",
        ylab="t",zlab="l(x,t)");
  persp(x,tp,hp,theta=60,phi=45,
        xlim=c(xlp[np],xu),ylim=c(tp[1],tp[np]),xlab="x",
        ylab="t",zlab="h(x,t)");
  persp(x,tp,pp,theta=60,phi=45,
        xlim=c(xlp[np],xu),ylim=c(tp[1],tp[np]),xlab="x",
        ylab="t",zlab="p(x,t)");
  persp(x,tp,qp,theta=60,phi=45,
        xlim=c(xlp[np],xu),ylim=c(tp[1],tp[np]),xlab="x",
        ylab="t",zlab="q(x,t)");
  persp(x,tp,mp,theta=60,phi=45,
        xlim=c(xlp[np],xu),ylim=c(tp[1],tp[np]),xlab="x",
        ylab="t",zlab="m(x,t)");
  persp(x,tp,Np,theta=60,phi=45,
        xlim=c(xlp[np],xu),ylim=c(tp[1],tp[np]),xlab="x",
        ylab="t",zlab="N(x,t)");
```

- The movement of x_l and the associated $\dfrac{dx_l}{dt}$ are plotted against t.

```
#
# Boundary position
  mon=60*60*24*30;
  mon=1;
```

```
   plot(tp/mon,xlp,xlab="t",ylab="xl");
     lines(tp/mon,xlp,type="l",lwd=2);
   points(tp/mon,xlp,pch="o",lwd=2);
   plot(tp[2:np]/mon,dxdtp,xlab="t",ylab="dru/dt");
     lines(tp[2:np]/mon,dxdtp,type="l",lwd=2);
   points(tp[2:np]/mon,dxdtp,pch="o",lwd=2);
```

`tp[2:np]` is used for t since the derivative array `dxdtp` does not include a value at $t = t_0 = 0$. The scaling of t can be in months rather than seconds with `mon`.

- The number of calls to the ODE/MOL routine, `pde_1a`, is displayed as a measure of the computational effort required to compute the complete solution.

```
#
# Calls to ODE routine
  cat(sprintf("\n\n  ncall = %3d\n",ncall));
```

This completes the discussion of the main program in Listing 5.1. The ODE/MOL routine `pde_1a` called by `ODE` in the main program is considered next.

5.2.2   ODE/MOL Routine

ODE/MOL routine `pde_1a` is listed next.

Listing 5.2: ODE/MOL routine pde_1a for Eqs. (5.1) and (5.2)

```
  pde_1a=function(t,u,parms){
#
# Function pde_1a computes the t derivative
# vectors of l(x,t),h(x,t),p(x,t),q(x,t),
# m(x,t),N(x,t)
#
# One vector to six vectors
  l=rep(0,nx);h=rep(0,nx);
  p=rep(0,nx);q=rep(0,nx);
  m=rep(0,nx);N=rep(0,nx);
  for(i in 1:nx){
    l[i]=u[i];
    h[i]=u[i+nx];
```
                                                              (Continued)

Listing 5.2 (Continued): ODE/MOL routine pde_1a for Eqs. (5.1) and (5.2)

```
    p[i]=u[i+2*nx];
    q[i]=u[i+3*nx];
    m[i]=u[i+4*nx];
    N[i]=u[i+5*nx];
  }
#
# lx,hx,px,qx,mx,Nx
  lx=rep(0,nx);hx=rep(0,nx);
  px=rep(0,nx);qx=rep(0,nx);
  mx=rep(0,nx);Nx=rep(0,nx);
  tablel=splinefun(x,l);lx=tablel(x,deriv=1);
  tableh=splinefun(x,h);hx=tableh(x,deriv=1);
  tablep=splinefun(x,p);px=tablep(x,deriv=1);
  tableq=splinefun(x,q);qx=tableq(x,deriv=1);
  tablem=splinefun(x,m);mx=tablem(x,deriv=1);
  tableN=splinefun(x,N);Nx=tableN(x,deriv=1);
#
# BCs
  lx[1]=-(k_l/D_l)*(l0-l[1]);lx[nx]=0;
  hx[1]=-(k_h/D_h)*(h0-h[1]);hx[nx]=0;
  px[1]=-(k_p/D_p)*(p0-p[1]);px[nx]=0;
  qx[1]=-(k_q/D_q)*(q0-q[1]);qx[nx]=0;
  mx[1]=-(k_m/D_m)*(m0-m[1]);mx[nx]=0;
  Nx[1]=-(k_N/D_N)*(N0-N[1]);Nx[nx]=0;
#
# lxx,hxx,pxx,qxx,mxx,Nxx
  lxx=rep(0,nx);hxx=rep(0,nx);
  pxx=rep(0,nx);qxx=rep(0,nx);
  mxx=rep(0,nx);Nxx=rep(0,nx);
  tablelx=splinefun(x,lx);lxx=tablelx(x,deriv=1);
  tablehx=splinefun(x,hx);hxx=tablehx(x,deriv=1);
  tablepx=splinefun(x,px);pxx=tablepx(x,deriv=1);
  tableqx=splinefun(x,qx);qxx=tableqx(x,deriv=1);
  tablemx=splinefun(x,mx);mxx=tablemx(x,deriv=1);
  tableNx=splinefun(x,Nx);Nxx=tableNx(x,deriv=1);
#
# PDEs
  lt=rep(0,nx);ht=rep(0,nx);
  pt=rep(0,nx);qt=rep(0,nx);
  mt=rep(0,nx);Nt=rep(0,nx);
```

(Continued)

Listing 5.2 (Continued): ODE/MOL routine pde_1a for Eqs. (5.1) and (5.2)

```
  for(i in 1:nx){
#
#    Product functions
     lm=l[i]/(r_l3+l[i])*m[i];
     hN=h[i]/(r_h3+h[i])*N[i];
#
#    l_t
     lt[i]=D_l*lxx[i]-
       r_l1*lm-r_l2*l[i];
#
#    h_t
     ht[i]=D_h*hxx[i]-
       r_h1*hN-r_h2*h[i];
#
#    p_t
     pt[i]=D_p*pxx[i]+
       r_p1*lm-r_p2*p[i];
#
#    q_t
     qt[i]=D_q*qxx[i]+
       r_q1*lm-r_q2*q[i];
#
#    m_t
     mt[i]=D_m*mxx[i]-
       r_m1*(mx[i]*lx[i]+m[i]*lxx[i])-
       r_m2*lm+r_m3*hN-r_m4*m[i]+
       r_m5*l[i];
#
#    N_t
     Nt[i]=D_N*Nxx[i]+
       r_N1*lm-r_N2*hN+
       r_N3*l[i];
  }
#
# Six vectors to one vector
  ut=rep(0,6*nx);
  for(i in 1:nx){
    ut[i]     =lt[i];
    ut[i+nx]   =ht[i];
    ut[i+2*nx]=pt[i];
```

(*Continued*)

Listing 5.2 (Continued): ODE/MOL routine **pde_1a** for Eqs. (5.1) and (5.2)

```
    ut[i+3*nx]=qt[i];
    ut[i+4*nx]=mt[i];
    ut[i+5*nx]=Nt[i];
  }
#
# Increment calls to pde_1a
  ncall <<- ncall+1;
#
# Return derivative vector
  return(list(c(ut)));
  }
```

We can note the following details about pde_1a in Listing 5.2.

- The function is defined.

```
    pde_1a=function(t,u,parms){
#
# Function pde_1a computes the t derivative
# vectors of l(x,t),h(x,t),p(x,t),q(x,t),
# m(x,t),N(x,t)
```

t is the current value of t in Eqs. (5.1). u the 156-vector of ODE/MOL dependent variables. parm is an argument to pass parameters to pde_1a (unused, but required in the argument list). The arguments must be listed in the order stated to properly interface with ode called in the main program of Listing 5.1. The derivative vector of the LHS of Eqs. (5.1) is calculated and returned to ode as explained subsequently.

- u is placed in six vectors, l,h,p,q,m,N, to facilitate the programming of Eqs. (5.1).

```
#
# One vector to six vectors
    l=rep(0,nx);h=rep(0,nx);
    p=rep(0,nx);q=rep(0,nx);
    m=rep(0,nx);N=rep(0,nx);
    for(i in 1:nx){
      l[i]=u[i];
      h[i]=u[i+nx];
      p[i]=u[i+2*nx];
      q[i]=u[i+3*nx];
```

```
    m[i]=u[i+4*nx];
    N[i]=u[i+5*nx];
}
```

- The first derivatives $\dfrac{\partial \ell}{\partial x}, \dfrac{\partial h}{\partial x}, \dfrac{\partial p}{\partial x}, \dfrac{\partial q}{\partial x}, \dfrac{\partial m}{\partial x}, \dfrac{\partial N}{\partial x}$, are computed by the spline function `splinefun`.

```
#
# lx,hx,px,qx,mx,Nx
  lx=rep(0,nx);hx=rep(0,nx);
  px=rep(0,nx);qx=rep(0,nx);
  mx=rep(0,nx);Nx=rep(0,nx);
  tablel=splinefun(x,l);lx=tablel(x,deriv=1);
  tableh=splinefun(x,h);hx=tableh(x,deriv=1);
  tablep=splinefun(x,p);px=tablep(x,deriv=1);
  tableq=splinefun(x,q);qx=tableq(x,deriv=1);
  tablem=splinefun(x,m);mx=tablem(x,deriv=1);
  tableN=splinefun(x,N);Nx=tableN(x,deriv=1);
```

`splinefun` is used in two steps. The first step computes a table of spline coefficients for a numerical vector. For example, for the vector N, the table of coefficients is computed with `tableN=splinefun(x,N)`. The second step computes the spline approximation of the derivative $\dfrac{\partial N}{\partial x}$ from the table, `Nx=tableN(x,deriv=1)`. `deriv=1` specifies a first derivative (`deriv=0` specifies the vector that is differentiated, and `deriv=2` specifies a second derivative of the vector).

- The Robin and Neumann BCs (5.2) are implemented (the subscripts 1,nx correspond to $x = x_l, x_u$).

```
#
# BCs
  lx[1]=-(k_l/D_l)*(l0-l[1]);lx[nx]=0;
  hx[1]=-(k_h/D_h)*(h0-h[1]);hx[nx]=0;
  px[1]=-(k_p/D_p)*(p0-p[1]);px[nx]=0;
  qx[1]=-(k_q/D_q)*(q0-q[1]);qx[nx]=0;
  mx[1]=-(k_m/D_m)*(m0-m[1]);mx[nx]=0;
  Nx[1]=-(k_N/D_N)*(N0-N[1]);Nx[nx]=0;
```

- The second derivatives $\dfrac{\partial^2 \ell}{\partial x^2}, \dfrac{\partial^2 h}{\partial x^2}, \dfrac{\partial^2 p}{\partial x^2}, \dfrac{\partial^2 q}{\partial x^2}, \dfrac{\partial^2 m}{\partial x^2}, \dfrac{\partial^2 N}{\partial x^2}$, are computed by differentiating the first derivatives (successive or stagewise differentiation).

```
#
# lxx,hxx,pxx,qxx,mxx,Nxx
  lxx=rep(0,nx);hxx=rep(0,nx);
  pxx=rep(0,nx);qxx=rep(0,nx);
  mxx=rep(0,nx);Nxx=rep(0,nx);
  tablelx=splinefun(x,lx);lxx=tablelx(x,deriv=1);
  tablehx=splinefun(x,hx);hxx=tablehx(x,deriv=1);
  tablepx=splinefun(x,px);pxx=tablepx(x,deriv=1);
  tableqx=splinefun(x,qx);qxx=tableqx(x,deriv=1);
  tablemx=splinefun(x,mx);mxx=tablemx(x,deriv=1);
  tableNx=splinefun(x,Nx);Nxx=tableNx(x,deriv=1);
```

`splinefun` is called with `deriv=1` for a first order differentiation of a first derivative.

- Eqs. (5.1) are programmed according to the following steps.

  - Vectors are defined for the LHS derivatives of Eqs. (5.1), $\dfrac{\partial \ell}{\partial t}$, $\dfrac{\partial h}{\partial t}$, $\dfrac{\partial p}{\partial t}$, $\dfrac{\partial q}{\partial t}$, $\dfrac{\partial m}{\partial t}$, $\dfrac{\partial N}{\partial t}$

```
#
# PDEs
  lt=rep(0,nx);ht=rep(0,nx);
  pt=rep(0,nx);qt=rep(0,nx);
  mt=rep(0,nx);Nt=rep(0,nx);
```

  - The products $\dfrac{\ell m}{r_{l3} + \ell}$, $\dfrac{h}{r_{h3} + h}$ are computed at a particular x (with a `for`) for use in the subsequent programming of Eqs. (5.1).

```
  for(i in 1:nx){
#
#    Product functions
     lm=l[i]/(r_l3+l[i])*m[i];
     hN=h[i]/(r_h3+h[i])*N[i];
```

  - Eq. (5.1-1) is programmed in the MOL format.

```
#
#    l_t
     lt[i]=D_l*lxx[i]-
       r_l1*lm-r_l2*l[i];
```

The correspondence of the programming and the mathematical statement of the PDE are an important feature of the MOL. The nonlinearity $\dfrac{\ell m}{r_{l3} + \ell}$ is easily included (`lm`).

  - Eq. (5.1-2) is programmed.

```
#
#   h_t
    ht[i]=D_h*hxx[i]-
        r_h1*hN-r_h2*h[i];
```

- Eq. (5.1-3) is programmed.

```
#
#   p_t
    pt[i]=D_p*pxx[i]+
        r_p1*lm-r_p2*p[i];
```

- Eq. (5.1-4) is programmed.

```
#
#   q_t
    qt[i]=D_q*qxx[i]+
        r_q1*lm-r_q2*q[i];
```

- Eq. (5.1-5) is programmed.

```
#
#   m_t
    mt[i]=D_m*mxx[i]-
        r_m1*(mx[i]*lx[i]+m[i]*lxx[i])-
        r_m2*lm+r_m3*hN-r_m4*m[i]+
        r_m5*l[i];
```

The chemotaxis diffusion term $r_{m1}(\dfrac{\partial m}{\partial x}\dfrac{\partial \ell}{\partial x} + m\dfrac{\partial^2 \ell}{\partial x^2})$ is easily included. Both nonlinear product functions are also included (with coefficients r_{m2}, r_{m3}).

- Eq. (5.1-6) is programmed.

```
#
#   N_t
    Nt[i]=D_N*Nxx[i]+
        r_N1*lm-r_N2*hN+
        r_N3*l[i];
  }
```

The final } concludes the **for** in x.

- The six vectors lt,ht,pt,qt,mt,Nt are placed in a single derivative vector ut of length $6(26) = 156$ to return to ode1.

```
#
# Six vectors to one vector
  ut=rep(0,6*nx);
  for(i in 1:nx){
    ut[i]      =lt[i];
    ut[i+nx]   =ht[i];
```

```
          ut[i+2*nx]=pt[i];
          ut[i+3*nx]=qt[i];
          ut[i+4*nx]=mt[i];
          ut[i+5*nx]=Nt[i];
      }
```

- The counter for the calls to **pde_1a** is incremented and returned to the main program of Listing 5.1 with <<-.

```
#
# Increment calls to pde_1a
    ncall <<- ncall+1;
```

- **ut** is returned to **ode** as a list (required by **ode**). **c** is the R vector utility.

```
#
# Return derivative vector
    return(list(c(ut)));
    }
```

The final } concludes **pde_1a**.

This concludes the discussion of **pde_1a** in Listing 5.2. The output of the main program and subordinate routine of Listings 5.1 and 5.2 is considered next.

5.2.3 Model Output

We can note the following details about this output (Table 5.2).

- The dimensions of the solution matrix **out** are $nout \times 6nx + 1 = 11 \times 6(26) + 1 = 157$ (11 values of t, and for each t, 26 values of x).

  The offset $+1$ results from the value of t as the first element in each of the $nout = 11$ solution vectors. These same values of t are in **tout**.

- ICs (5.3) $(t = 0)$ are verified for the homogeneous (zero) case.

- The output is for $x = 0, 0.004/25, ..., 0.004$ as programmed in Listing 5.1 (26 values of x at each value of t with every fifth value in x displayed).

- The output is for $t = 0, 1.0 \times 10^7/10, ..., 1.0 \times 10^7$ as programmed in Listing 5.1.

- l(x=xu=4.0e-03,t=1.0e+07)=9.772e-01

  h(x=xu=4.0e-03,t=1.0e+07)=2.933e+00

  which are close to the bloodstream concentrations l0=1, h0=3 set in Listing 5.1.

- The computational effort is manageable, `ncall` = 13329 (that is, this number appears to be large, but the solution is computed in reasonable time).

The graphical output is in Figures 5.1.

In Figures 5.1-1 through 5.1-6, the variation of the solutions in x is small because the EIL is thin (40 μm).

For `ncase=1`, the EIL inner boundary remains at $x_l = 0$ as programmed in Listing 5.1. This case is worth checking since $x_l \neq 0$ would indicate a programming error.

For `ncase=1`, $dx_l/dt = 0$. This case is worth checking since $dx_l/dt \neq 0$ would indicate a programming error.

The output for `ncase=2` follows (Table 5.3).

We can note the following details about this output.

- The dimensions of the solution matrix `out` are $nout \times 6nx + 1 = 11 \times 6(26) + 1 = 157$ (11 values of t, and for each t, 26 values of x).

  The offset $+1$ results from the value of t as the first element in each of the $nout = 11$ solution vectors. These same values of t are in `tout`,

- ICs (5.3) ($t = 0$) are verified for the homogeneous (zero) case.

- The output is for $t = 0, 1.0 \times 10^7/10, ..., 1.0 \times 10^7$ as programmed in Listing 5.1.

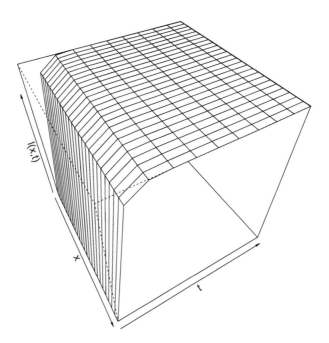

FIGURE 5.1-1
Numerical solution $\ell(x, t)$ from Eq. (5.1-1).

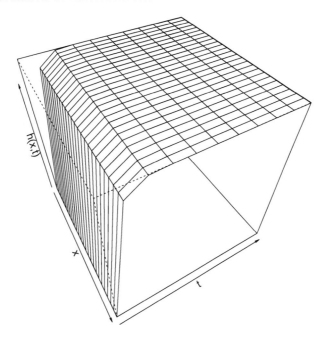

FIGURE 5.1-2
Numerical solution $h(x,t)$ from Eq. (5.1-2).

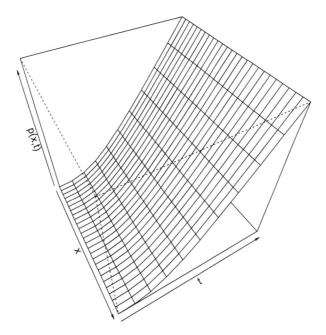

FIGURE 5.1-3
Numerical solution $p(x,t)$ from Eq. (5.1-3).

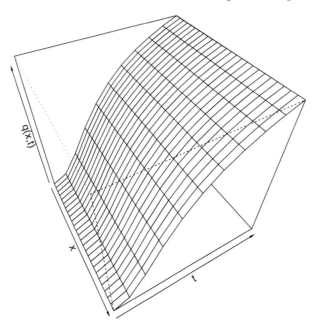

FIGURE 5.1-4
Numerical solution $q(x,t)$ from Eq. (5.1-4).

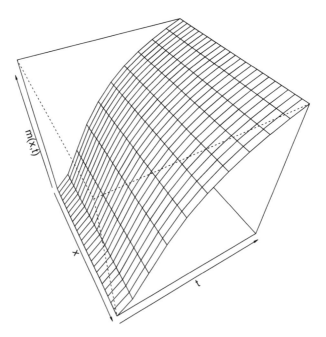

FIGURE 5.1-5
Numerical solution $m(x,t)$ from Eq. (5.1-5).

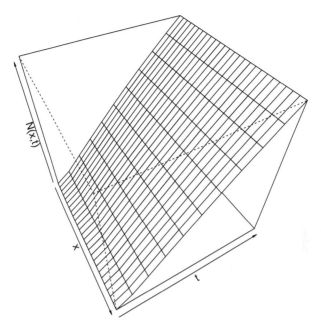

FIGURE 5.1-6
Numerical solution $N(x,t)$ from Eq. (5.1-6).

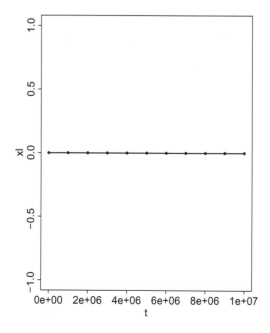

FIGURE 5.1-7
x_l from Eq. (5.2-13).

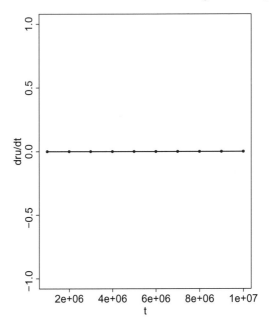

FIGURE 5.1-8
dx_l/dt for `ncase=1`.

- The output is for $x = 0, ..., 0.004$ at $t = 0$ to $x = -0.01, ..., 0.004$ at $t = 1.0 \times 10^7$ as programmed in Listing 5.1 (26 values of x at each value of t with every 5th value in x displayed).

  This change in x_l from 0 to -0.01 follows from $\dfrac{dx_l}{dt} = -1.0 \times 10^{-9}$ (`ncase=2` in Listing 5.1) and $\dfrac{dx_l}{dt} dt = (-1.0 \times 10^{-9})(1.0 \times 10^7) = -0.01$.

- `l(x=xl=-1.0e-02,t=1.0e+07)=9.328e-01`

  `h(x=xl=-1.0e-02,t=1.0e+07)=2.806e+00`

  which are close to the bloodstream concentrations `l0=1`, `h0=3` set in Listing 5.1. Also, these values are close to the `ncase=1` values (compare Tables 5.2 and 5.3) and the solutions for `ncase=1,2` are similar, even with the movement of the boundary at $x = x_l$ from 0 to -0.01.

- The computational effort is still manageable, `ncall = 18484`.

The graphical output is in Figures 5.2. Since the solutions for $\ell(x,t)$ to $N(x,t)$ are similar to those for `ncase=1` (Figures 5.1-1 through 5.1-6), the graphical output is limited to $m(x,t)$ to conserve space (compare Figure 5.2-5 with 5.1-5).

TABLE 5.2
Abbreviated numerical output for `ncase=1`

t	x	l(x,t)
t	x	h(x,t)
t	x	p(x,t)
t	x	q(x,t)
t	x	m(x,t)
t	x	N(x,t)
0.000e+00	0.000e+00	0.000e+00
0.000e+00	0.000e+00	0.000e+00
0.000e+00	0.000e+00	0.000e+00
0.000e+00	0.000e+00	0.000e+00
0.000e+00	0.000e+00	0.000e+00
0.000e+00	0.000e+00	0.000e+00
.		.
.		.
.		.

Output for x = 8.0e-04
to 3.2e-04 removed

.		.
.		.
.		.
0.000e+00	4.000e-03	0.000e+00
0.000e+00	4.000e-03	0.000e+00
0.000e+00	4.000e-03	0.000e+00
0.000e+00	4.000e-03	0.000e+00
0.000e+00	4.000e-03	0.000e+00
0.000e+00	4.000e-03	0.000e+00
.		.
.		.
.		.

Output for t = 1.0e+06
to 9.0+06 removed

.		.
.		.
.		.
t	x	l(x,t)
t	x	h(x,t)
t	x	p(x,t)
t	x	q(x,t)
t	x	m(x,t)
t	x	N(x,t)

(Continued)

TABLE 5.2 (*Continued*)
Abbreviated numerical output for `ncase=1`

1.000e+07	0.000e+00	9.772e-01
1.000e+07	0.000e+00	2.933e+00
1.000e+07	0.000e+00	4.319e-02
1.000e+07	0.000e+00	3.196e-03
1.000e+07	0.000e+00	1.655e-01
1.000e+07	0.000e+00	4.162e-01

```
               .               .
               .               .
               .               .

        Output for x = 8.0e-04
           to 3.2e-04 removed

               .               .
               .               .
               .               .
```

1.000e+07	4.000e-03	9.772e-01
1.000e+07	4.000e-03	2.933e+00
1.000e+07	4.000e-03	4.319e-02
1.000e+07	4.000e-03	3.203e-03
1.000e+07	4.000e-03	1.655e-01
1.000e+07	4.000e-03	4.162e-01

```
ncall = 13329
```

TABLE 5.3
Abbreviated numerical output for `ncase=2`

t	x	l(x,t)
t	x	h(x,t)
t	x	p(x,t)
t	x	q(x,t)
t	x	m(x,t)
t	x	N(x,t)
0.000e+00	0.000e+00	0.000e+00
0.000e+00	0.000e+00	0.000e+00
0.000e+00	0.000e+00	0.000e+00
0.000e+00	0.000e+00	0.000e+00
0.000e+00	0.000e+00	0.000e+00
0.000e+00	0.000e+00	0.000e+00

```
               .               .
               .               .
               .               .
```

(Continued)

TABLE 5.3 (*Continued*)
Abbreviated numerical output for `ncase=2`

```
        Output for x = 8.0e-04
           to 3.2e-04 removed
               .                  .

               .                  .

               .                  .
0.000e+00    4.000e-03    0.000e+00
0.000e+00    4.000e-03    0.000e+00
0.000e+00    4.000e-03    0.000e+00
0.000e+00    4.000e-03    0.000e+00
0.000e+00    4.000e-03    0.000e+00
0.000e+00    4.000e-03    0.000e+00
               .                  .

            .     .                  .

               .                  .
        Output for t = 1.0e+06
           to 9.0+06 removed
               .                  .

               .                  .

               .                  .
1.000e+07    -1.000e-02    9.328e-01
1.000e+07    -1.000e-02    2.806e+00
1.000e+07    -1.000e-02    4.458e-02
1.000e+07    -1.000e-02    9.502e-03
1.000e+07    -1.000e-02    1.727e-01
1.000e+07    -1.000e-02    4.086e-01
               .                  .

               .                  .

               .                  .
        Output for t = 1.0e+06
           to 9.0+06 removed
               .                  .

               .                  .

               .                  .
1.000e+07    4.000e-03    9.323e-01
1.000e+07    4.000e-03    2.804e+00
1.000e+07    4.000e-03    4.458e-02
1.000e+07    4.000e-03    9.573e-03
1.000e+07    4.000e-03    1.727e-01
1.000e+07    4.000e-03    4.086e-01

ncall = 18484
```

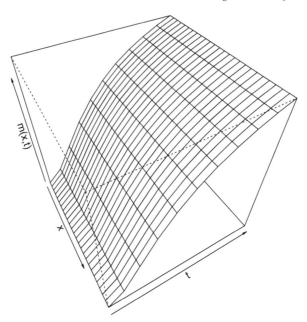

FIGURE 5.2-5
Solution for $m(x,t)$, `ncase=2`.

$x_l(x,t)$ varies linearly from $x_l(t = 0) = 0$ to $x_l(t = 1.0 \times 10^7) = -0.01$ (Table 5.4) reflecting the movement of the boundary for `ncase=2`.

$$\frac{dx_l}{dt} = -1.0 \times 10^{-9} \text{ for } \texttt{ncase=2}.$$

The movement of the boundary ($0 \leq x_l \leq 0.004$ for `ncase=1` to $-0.01 \leq x_l \leq 0.004$ for `ncase=2` corresponds to a $0.01/0.004 \times 100 = 250\%$ increase in x_l.

The output for `ncase=3` follows (Table 5.4).

We can note the following details about this output.

- The dimensions of the solution matrix `out` are $nout \times 6nx + 1 = 11 \times 6(26) + 1 = 157$ (11 values of t, and for each t, 26 values of x).

  The offset $+1$ results from the value of t as the first element in each of the $nout = 11$ solution vectors. These same values of t are in `tout`,

- ICs (5.3) ($t = 0$) are verified for the homogeneous (zero) case.

- The output is for $t = 0, 1.0 \times 10^7/10, ..., 1.0 \times 10^7$ as programmed in Listing 5.1.

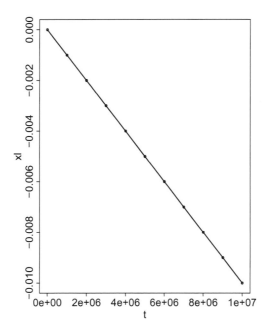

FIGURE 5.2-7
Variation of $x_l(x,t)$ with t.

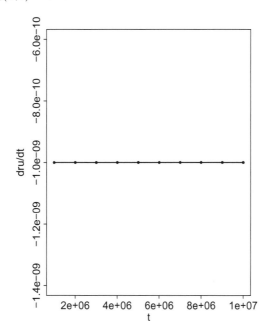

FIGURE 5.2-8
Variation of $dx_l(x,t)/dt$ with t.

- The output is for $x = 0, ..., 0.004$ at $t = 0$ to $x = -0.02286, ..., 0.004$ at $t = 1.0 \times 10^7$ as programmed in Listing 5.1 (26 values of x at each value of t with every 5th value in x displayed).

  This change in x_l from 0 to -0.02286 follows from $\dfrac{dx_l}{dt} == -k_{x_l} N(x = x_l, t)$ of Eq. (5.2-13) (with ncase=3 in Listing 5.1).

- x_l and $\dfrac{dx_l}{dt}$ vary with t as reflected in Figure 5.3-7 and 8 (given below).

TABLE 5.4
Abbreviated numerical output for ncase=3

t	x	l(x,t)
t	x	h(x,t)
t	x	p(x,t)
t	x	q(x,t)
t	x	m(x,t)
t	x	N(x,t)
0.000e+00	0.000e+00	0.000e+00
0.000e+00	0.000e+00	0.000e+00
0.000e+00	0.000e+00	0.000e+00
0.000e+00	0.000e+00	0.000e+00
0.000e+00	0.000e+00	0.000e+00
0.000e+00	0.000e+00	0.000e+00

.
.
.

Output for x = 8.0e-04
to 3.2e-04 removed

.
.
.

0.000e+00	4.000e-03	0.000e+00
0.000e+00	4.000e-03	0.000e+00
0.000e+00	4.000e-03	0.000e+00
0.000e+00	4.000e-03	0.000e+00
0.000e+00	4.000e-03	0.000e+00
0.000e+00	4.000e-03	0.000e+00

.
.
.

(Continued)

TABLE 5.4 (*Continued*)
Abbreviated Numerical Output for `ncase=3`

```
        Output for t = 1.0e+06
           to 9.0+06 removed
                 .                    .

                 .                    .

                 .                    .
    1.000e+07   -2.286e-02    9.018e-01
    1.000e+07   -2.286e-02    2.719e+00
    1.000e+07   -2.286e-02    4.636e-02
    1.000e+07   -2.286e-02    1.442e-02
    1.000e+07   -2.286e-02    1.864e-01
    1.000e+07   -2.286e-02    4.078e-01

                 .                    .

                 .                    .

                 .                    .
        Output for t = 1.0e+06
           to 9.0+06 removed
                 .                    .

                 .                    .

                 .                    .
    1.000e+07    4.000e-03    9.002e-01
    1.000e+07    4.000e-03    2.714e+00
    1.000e+07    4.000e-03    4.636e-02
    1.000e+07    4.000e-03    1.465e-02
    1.000e+07    4.000e-03    1.864e-01
    1.000e+07    4.000e-03    4.078e-01

    ncall = 18074
```

- `l(x=xl=-0.02286,t=1.0e+07)=9.018e-01`

  `h(x=xl=-0.02286,t=1.0e+07)=2.719e+00`

  which depart from the bloodstream concentrations `l0=1`, `h0=3` set in Listing 5.1 as a consequence of Eq. (5.2-13). However, the solutions for $\ell(x,t)$ to $N(x,t)$ are similar to the solutions for `ncase=1,2` (Figures 5.1 and 5.2).

- The computational effort is still manageable, `ncall = 18074`.

The graphical output is in Figures 5.3. Since the solutions for $\ell(x,t)$ to $N(x,t)$ are similar to those for `ncase=1` (Figures 5.1-1 through 5.1-6) and `ncase=2`

(Figures 5.2-1 through 5.2-6), the graphical output is limited to $m(x,t)$ to conserve space (compare Figure 5.3-5 with Figures 5.1-5 and 5.2-5).

$x_l(x,t)$ varies from $x_l(t = 0) = 0$ to $x_l(t = 1.0 \times 10^7) = -0.02286$ (Table 5.4) reflecting the movement of the boundary for ncase=3.

$$\frac{dx_l}{dt} = -k_{x_l}N(x = x_l, t) \text{ (Eq. 5.2-13) for ncase=3.}$$

The movement of the boundary $0 \le x_l \le 0.004$ for ncase=1 to $-0.02286 \le x_l \le 0.004$ for ncase=3 corresponds to a $0.02286/0.004 \times 100 = 571.5\%$ increase in x_l.

This substantial increase in the EIL can be interrupted as a result of plaque formation at the EIL inner boundary. This presumes that PDEs (5.1) apply to the addition in x, that is $-0.02286 \le x \le 0$. If the properties of this added region are basically different, they can be altered in the ODE/MOL routine pde_1a (by changing the properties with x). Or the PDEs can be changed as a function of x. Thus, extensions of the model of Eqs. (5.1) through (5.3) for plaque formation can be considered and, in principle, implemented for further computer-based study.

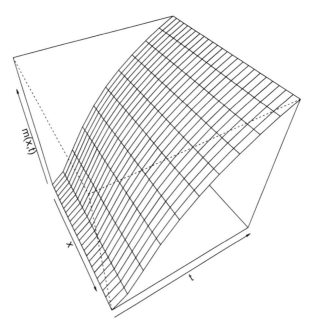

FIGURE 5.3-5
Solution for $m(x,t)$, ncase=3.

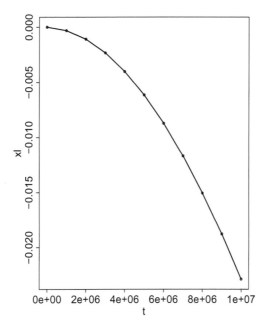

FIGURE 5.3-7
Variation of $x_l(x,t)$ with t.

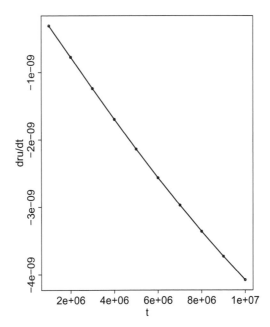

FIGURE 5.3-8
Variation of $dx_l(x,t)/dt$ with t.

5.3   Summary and Conclusions

Eqs. (5.1) through (5.3) constitute the PDE model for the movement of the EIL inner boundary according to a presumed equation of motion, Eq. (5.2-13). The R coding (programming) of the MOL is discussed in detail, and the output from the R routines is analyzed in detail.

The solutions to Eqs. (5.1) through (5.3) have the following distinguishing features: (1) small variation in x which is due to the thin EIL ($0 \le x \le 0.004$ at $t = 0$) and the linear Fick's diffusion and the nonlinear chemotaxis diffusion which tend to disperse the solutions in x, (2) smoothness even with nonlinear, coupling, product functions, e.g., $m(x,t)$ from Eq. (5.1-5) in Figures 5.1-5 through 5.3-5, and (3) little variation in the form of the solutions with the addition of the moving boundary at $x = x_l$ which is probably due to the l_0, h_0 forcing in BCs (5.2-1) and (5.2-3) as the solutions evolve in t; this explanation could be studied further by numerically examining the individual RHS and LHS terms of Eq. (5.1).

The intent of the model is to provide a quantitative framework for the investigation of atherosclerosis on modest computers. Hopefully, it will provide a small step toward an improved understanding of atherosclerosis. The basic concepts of the molecular and cellular mechanisms represented by the PDE model are discussed in detail in [1–4], and these references are gratefully acknowledged.

References

1. Chalmers, A.D., A. Cohen, C.A. Bursill and M.R. Myerscough (2015), Bifurcation and dynamics in a mathematical model of early atherosclerosis, *J. Math. Biol.*, **71**, 1451–1480.

2. Chalmers, A.D., C.A. Bursill and M.R. Myerscough (2017), Nonlinear dynamics of early atherosclerosis plaque formation may determine the effect of high density lipoproteins in plaque regressions, *PLoS One*, **12**, no. 11.

3. Hao, W., and A. Friedman (2014), The LDL-HDL profile determines the risk of atherosclerosis: A mathematical model, *PLoS One*, **9**, no. 3.

4. Khatib, N.E., S. Genieys, B. Kazmierczak and V. Volpert (2012), Reaction-diffusion model of atherosclerosis development, *J. Math. Bio*, **65**, 349–374.

5. Schiesser, W.E. (2019), *PDE Models for Atherosclerosis: Computer Implementation in R*, Morgan and Claypool, San Rafael, CA, USA.

6. Soetaert, K., J. Cash, and F. Mazzia (2012), *Solving Differential Equations in R*, Springer-Verlag, Heidelberg, Germany.

Appendix A1

Test of Spline Regridding

In Chapter 2, an algorithm for a partial differential equation (PDE) with a moving boundary is developed and applied to the diffusion equation in 1D cylindrical coordinates (Listings 2.1 and 2.2). Chapter 3 includes a discussion of the output of the routines in Chapter 2 for three cases. In this appendix, the moving boundary algorithm is tested with a Gaussian function.

First, the spline interpolation/extrapolation is applied to a Gaussian function with a fixed boundary.

A1.1 Fixed Boundary Analysis

Listing A1.1: Main program for fixed boundary, $r_u = 1$

```
#
# Gaussian test function
#
# Delete previous workspaces
  rm(list=ls(all=TRUE))
#
# Access functions for numerical solution
  setwd("f:/mbpde/gauss1");
#
# Grid (in r)
  n=21;rl=0;ru=1;
  r=seq(from=rl,to=ru,by=(ru-rl)/(n-1));
#
# Gaussian function
  u=rep(0,n);
  for(ir in 1:n){
    u[ir]=exp(-10*r[ir]^2);
  }
```

(Continued)

Listing A1.1 (Continued): Main program for fixed boundary, $r_u = 1$

```
#
# us
  tabler=splinefun(r,u);
  us=tabler(r,deriv=0);
#
# Error norm
  erru=0;
  for(ir in 1:n){
    erru=erru+(us[ir]-u[ir])^2;
  }
  erru=sqrt(erru/n);
  cat(sprintf("\n erru = %10.3e\n",erru));
#
# Plot function
  plot(r,u,xlab="r",ylab="u(r)");
   lines(r,u,type="l",lwd=2);
  points(r,us,pch="o",lwd=2);
#
# First derivative
  ur=rep(0,n);
  for(ir in 1:n){
    ur[ir]=(-2*10*r[ir])*exp(-10*r[ir]^2);
  }
#
# urs
  tabler=splinefun(r,us);
  urs=tabler(r,deriv=1);
#
# Error norm
  errur=0;
  for(ir in 1:n){
    errur=errur+(urs[ir]-ur[ir])^2;
  }
  errur=sqrt(errur/n);
  cat(sprintf("\n errur = %10.3e\n",errur));
#
# Plot first derivative
  plot(r,ur,xlab="r",ylab="ur(r)");
   lines(r,ur,type="l",lwd=2);
  points(r,urs,pch="o",lwd=2);
```

(Continued)

Listing A1.1 (Continued): Main program for fixed boundary, $r_u = 1$

```
#
# Second derivative (direct)
  urr=rep(0,n);
  for(ir in 1:n){
    urr[ir]=(-2*10*r[ir])^2*exp(-10*r[ir]^2)-
            (2*10)*exp(-10*r[ir]^2);
  }
#
# urrs
  tabler=splinefun(r,us);
  urrs=tabler(r,deriv=2);
#
# Error norm
  errurr=0;
  for(ir in 1:n){
    errurr=errurr+(urrs[ir]-urr[ir])^2;
  }
  errurr=sqrt(errurr/n);
  cat(sprintf("\n errurr = %10.3e\n",errurr));
#
# Plot second derivative
  plot(r,urr,xlab="r",ylab="urr(r)");
   lines(r,urr,type="l",lwd=2);
  points(r,urrs,pch="o",lwd=2);
#
# Second derivative (stagewise)
#
# urs
  tabler=splinefun(r,us);
  urs=tabler(r,deriv=1);
#
# urrs
  tabler=splinefun(r,urs);
  urrs=tabler(r,deriv=1);
#
# Error norm
  errurr=0;
  for(ir in 1:n){
    errurr=errurr+(urrs[ir]-urr[ir])^2;
  }
```

(Continued)

Listing A1.1 (Continued): Main program for fixed boundary, $r_u = 1$

```
errurr=sqrt(errurr/n);
cat(sprintf("\n errurr = %10.3e\n",errurr));
#
# Plot second derivative
plot(r,urr,xlab="r",ylab="urr(r)");
  lines(r,urr,type="l",lwd=2);
points(r,urrs,pch="o",lwd=2);
```

We can note the following details about Listing A1.1.

- Previous workspaces are deleted.

```
#
# Gaussian test function
#
# Delete previous workspaces
  rm(list=ls(all=TRUE))
```

- The directory with the R routines is designated. Note that setwd (set working directory) uses / rather than the usual \.

```
#
# Access functions for numerical solution
  setwd("f:/mbpde/gauss");
```

- A base grid in the spatial variable r is defined, $r_l = 0 \le r \le r_u = 1$, with 21 points.

```
#
# Grid (in r)
  n=21;rl=0;ru=1;
  r=seq(from=rl,to=ru,by=(ru-rl)/(n-1));
```

- A Gaussian function,

$$u(r) = e^{-10r^2}, \tag{A1.1}$$

is defined on the grid. The final objective of this analysis is to compute this function and its first and second derivatives on an expanded grid by increasing r_u, but initially, $r_u = 1$ is used to test the coding in Listing A1.1.

```
#
# Gaussian function
  u=rep(0,n);
  for(ir in 1:n){
    u[ir]=exp(-10*r[ir]^2);
  }
```

- $u(r)$ is approximated with a spline in two steps: (1) a table of spline coefficients, `tabler`, is defined with `splinefun`, a function in the basic R and (2) the spline approximation of $u(r)$, `us` $= u_s(r)$, is computed by interpolation within in the table `tabler` by specifying `deriv=0`.

```
#
# us
  tabler=splinefun(r,u);
  us=tabler(r,deriv=0);
```

- The difference between $u(r)$ and its spline approximation $u_s(r)$ is assessed with a norm $err(r(u))$.

$$err(u(r)) = \left(\frac{1}{n} \sum_1^n (u_s(r) - u(r))^2 \right)^{1/2}. \qquad (A1.2)$$

```
#
# Error norm
  erru=0;
  for(ir in 1:n){
    erru=erru+(us[ir]-u[ir])^2;
  }
  erru=sqrt(erru/n);
  cat(sprintf("\n erru = %10.3e\n",erru));
```

The norm is then displayed.

- $u(r)$ and $u_s(r)$ are superimposed on a plot to facilitate a comparison of the two.

```
#
# Plot function
  plot(r,u,xlab="r",ylab="u(r)");
  lines(r,u,type="l",lwd=2);
  points(r,us,pch="o",lwd=2);
```

The resulting plot is in Figure A1.1-1 that follows.

- The first derivative, $\dfrac{du_s(r)}{dr}$, computed from $u_s(r)$, is evaluated, first by computing the exact derivative.

$$\frac{du(r)}{dr} = (-2(10)r)e^{-10r^2}. \qquad (A1.3)$$

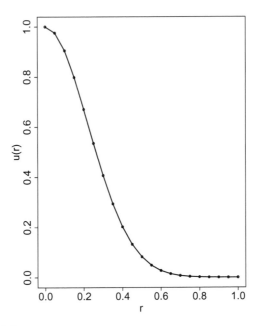

FIGURE A1.1-1
Comparison of fixed boundary exact and spline solutions: `ru=1`, `n=21`.

```
#
# First derivative
  ur=rep(0,n);
  for(ir in 1:n){
    ur[ir]=(-2*10*r[ir])*exp(-10*r[ir]^2);
  }
```

- $\dfrac{du_s(r)}{dr} =$ `urs` is computed with a spline.

```
#
# urs
  tabler=splinefun(r,us);
  urs=tabler(r,deriv=1);
```

- The difference between $\dfrac{du(r)}{dr}$ and its spline approximation $\dfrac{du_s(r)}{dr}$ is assessed with a norm $err\left(\dfrac{dr(u)}{dr}\right) =$ `errur`.

$$err\left(\frac{du(r)}{dr}\right) = \left(\frac{1}{n}\sum_{1}^{n}\left(\frac{du_s(r)}{dr} - \frac{du(r)}{dr}\right)^2\right)^{1/2}. \qquad (A1.4)$$

```
#
# Error norm
  errur=0;
  for(ir in 1:n){
    errur=errur+(urs[ir]-ur[ir])^2;
  }
  errur=sqrt(errur/n);
  cat(sprintf("\n errur = %10.3e\n",errur));
```

- $\dfrac{du(r)}{dr}$ and $\dfrac{du_s(r)}{dr}$ are superimposed on a plot to facilitate a comparison of the two.

```
#
# Plot first derivative
  plot(r,ur,xlab="r",ylab="ur(r)");
  lines(r,ur,type="l",lwd=2);
  points(r,urs,pch="o",lwd=2);
```

The resulting plot is in Figure A1.1-2 that follows.

- The second derivative, $\dfrac{d^2 u_s(r)}{dr^2}$, computed from $u_s(r)$, is evaluated,

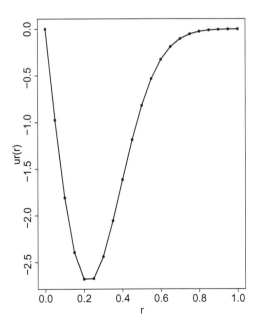

FIGURE A1.1-2
Comparison of fixed boundary exact and spline first derivatives: ru=1, n=21.

first by computing the exact derivative.

$$\frac{d^2u(r)}{dr^2} = (-2(10)r)^2 e^{-10r^2} - 2(10)e^{-10r^2}. \qquad (A1.5)$$

```
#
# Second derivative (direct)
  urr=rep(0,n);
  for(ir in 1:n){
    urr[ir]=(-2*10*r[ir])^2*exp(-10*r[ir]^2)-
            (2*10)*exp(-10*r[ir]^2);
  }
```

- The second derivative is computed directly (`deriv=2`) from u_s.

```
#
# urrs
  tabler=splinefun(r,us);
  urrs=tabler(r,deriv=2);
```

- The difference between $\dfrac{d^2u(r)}{dr^2}$ and its spline approximation $\dfrac{d^2u_s(r)}{dr^2}$ is assessed with a norm $err\left(\dfrac{d^2r(u)}{dr^2}\right) = $ `errurr`.

$$err\left(\frac{d^u(r)}{dr^2}\right) = \left(\frac{1}{n}\sum_1^n \left(\frac{d^2u_s(r)}{dr^2} - \frac{d^2u(r)}{dr^2}\right)^2\right)^{1/2}. \qquad (A1.6)$$

```
#
# Error norm
  errurr=0;
  for(ir in 1:n){
    errurr=errurr+(urrs[ir]-urr[ir])^2;
  }
  errurr=sqrt(errurr/n);
  cat(sprintf("\n errurr = %10.3e\n",errurr));
```

- $\dfrac{d^2u(r)}{dr^2}$ and $\dfrac{d^2u_s(r)}{dr^2}$ are superimposed on a plot to facilitate a comparison of the two.

```
#
# Plot second derivative
  plot(r,urr,xlab="r",ylab="urr(r)");
    lines(r,urr,type="l",lwd=2);
  points(r,urrs,pch="o",lwd=2);
```

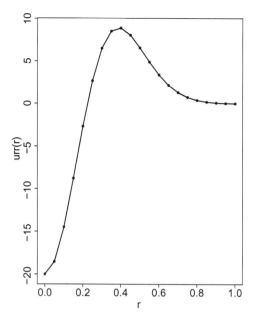

FIGURE A1.1-3
Comparison of fixed boundary exact and spline second derivatives: Direct differentiation, `ru=1`, `n=21`.

The resulting plot is in Figure A1.1-3.

- The second derivative is computed by successive differentiation (`deriv=1` twice) from u_s.

```
#
# Second derivative (stagewise)
#
# urs
   tabler=splinefun(r,us);
   urs=tabler(r,deriv=1);
#
# urrs
   tabler=splinefun(r,urs);
   urrs=tabler(r,deriv=1);
```

- The difference between $\dfrac{d^2 u(r)}{dr^2}$ and its spline approximation $\dfrac{d^2 u_s(r)}{dr^2}$ is assessed with a norm $err\left(\dfrac{d^2 r(u)}{dr^2}\right)$ = `errurr` of eq. (A1.6).

```
#
# Error norm
```

```
errurr=0;
for(ir in 1:n){
   errurr=errurr+(urrs[ir]-urr[ir])^2;
}
errurr=sqrt(errurr/n);
cat(sprintf("\n errurr = %10.3e\n",errurr));
```

- $\dfrac{d^2u(r)}{dr^2}$ and $\dfrac{d^2u_s(r)}{dr^2}$ are superimposed on a plot to facilitate a comparison of the two.

```
#
# Plot second derivative
  plot(r,urr,xlab="r",ylab="urr(r)");
   lines(r,urr,type="l",lwd=2);
  points(r,urrs,pch="o",lwd=2);
```

The resulting plot is in Figure A1.1-4 that follows.

This completes the programming of the Gaussian test function for a fixed boundary $r_u = 1$.

We can note the following details about the output.

- The spline returns the values of $u(r)$ on the predefined grid in r to full machine accuracy, `9.505e-20` (a property of splines).

- The spline differentiation gives first derivative values with an average error `6.373e-03/2.5*100 = 0.25%` (that has been scaled by the approximately largest value of the first derivative, `2.5`, from the vertical scale of Figure A1.1-2).

- The spline differentiation gives second derivative values with an average error `4.930e-01/10*100 = 4.9%` by direct differentiation (that has been scaled by the approximately largest value of the second derivative, `10`, from the vertical scale of Figure A1.1-3) and `3.862e-01/10*100 = 3.9%` by successive differentiation (that has been scaled by the approximately largest value of the second derivative, `10`, from the vertical scale of Figure A1.1-4).

Figures A1.1-1 through A1.1-4 indicate acceptable agreement between the exact and approximate values of the function and its first and second derivatives for the interval $r_l = 0 \leq r \leq r_u = 1$. Also, the accuracy of the approximate derivatives decreases with the order of the derivative.

The accuracy of the first and second derivatives is determined by the number of grid points in r. To investigate this aspect of the numerical solution and its derivatives, the preceding calculations are repeated with the number

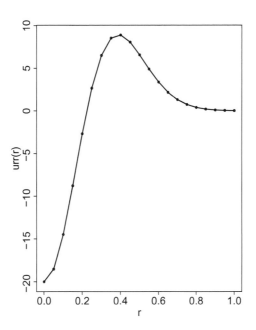

FIGURE A1.1-4
Comparison of fixed boundary exact and spline second derivatives: Successive
(stagewise) differentiation, `ru=1`, `n=21`.

of points increased from $n = 21$ to $n = 41$ (in Listing A1.1). The numerical
output follows.

The errors are reduced by increasing n from 21 to 41 (compare Tables A1.1
and A1.2).

The graphical output is in Figures A1.2-1 through A1.2-4.

Figures A1.2-1 through A1.2-4 are similar to Figures A1.1-1 through A1.1-4
through A1.4 but with the apparent increase in n.

TABLE A1.1
Numerical output, fixed boundary

erru =	9.505e-20
errur =	6.373e-03
errurr =	4.930e-01
errurr =	3.862e-01

TABLE A1.2
Numerical output, fixed boundary, n=41

erru =	6.773e-20
errur =	6.600e-04
rrurr =	1.016e-01
errurr =	7.410e-02

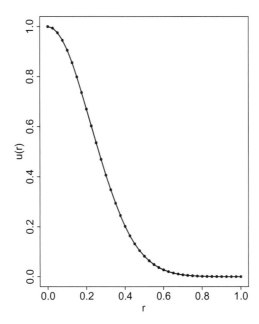

FIGURE A1.2-1
Comparison of fixed boundary exact and spline solutions: ru=1, n=41.

The preceding examples are for a fixed boundary $r_u = 1$ and the spline approximations constitute interpolation only. The case of a moving boundary with $r_u > 1$ is considered next, which includes spline extrapolation for the extended interval $r_l = 0 \le r \le r_u > 1$.

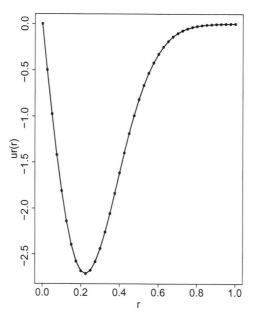

FIGURE A1.2-2
Comparison of fixed boundary exact and spline first derivatives: `ru=1, n=41`.

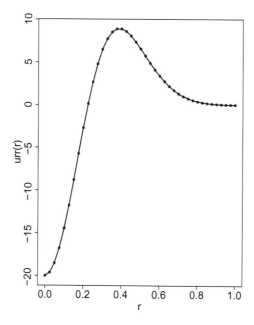

FIGURE A1.2-3
Comparison of fixed boundary exact and spline second derivatives: Direct differentiation, `ru=1, n=41`.

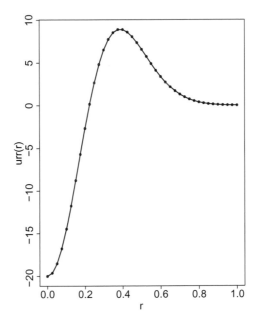

FIGURE A1.2-4
Comparison of fixed boundary exact and spline second derivatives: Successive
(stagewise) differentiation, ru=1, n=41.

A1.2   Moving Boundary Analysis

A R routine that includes the extended domain in r follows.

Listing A1.2: Main program for moving boundary, $r_u > 1$

```
#
# Gaussian test function
#
# Delete previous workspaces
  rm(list=ls(all=TRUE))
#
# Access functions for numerical solution
  setwd("f:/mbpde/gauss2");
#
# Grid (in r)
  n=21;rl=0;ru=1;
  r=seq(from=rl,to=ru,by=(ru-rl)/(n-1));
```

(Continued)

Listing A1.2 (Continued): Main program for moving boundary, $r_u > 1$

```
#
# Gaussian function
  u=rep(0,n);
  for(ir in 1:n){
    u[ir]=exp(-10*r[ir]^2);
  }
#
# us
  tabler=splinefun(r,u);
  ru=1.25;
  r=seq(from=rl,to=ru,by=(ru-rl)/(n-1));
  us=tabler(r,deriv=0);
  for(ir in 1:n){
    u[ir]=exp(-10*r[ir]^2);
  }
#
# Error norm
  erru=0;
  for(ir in 1:n){
    erru=erru+(us[ir]-u[ir])^2;
  }
  erru=sqrt(erru/n);
  cat(sprintf("\n erru = %10.3e\n",erru));
#
# Plot function
  plot(r,u,xlab="r",ylab="u(r)");
   lines(r,u,type="l",lwd=2);
  points(r,us,pch="o",lwd=2);
#
# First derivative
  ur=rep(0,n);
  for(ir in 1:n){
    ur[ir]=(-2*10*r[ir])*exp(-10*r[ir]^2);
  }
#
# urs
  tabler=splinefun(r,us);
  urs=tabler(r,deriv=1);
#
# Error norm
  errur=0;
```

(Continued)

Listing A1.2 (Continued): Main program for moving boundary, $r_u > 1$

```
  for(ir in 1:n){
    errur=errur+(urs[ir]-ur[ir])^2;
  }
  errur=sqrt(errur/n);
  cat(sprintf("\n errur = %10.3e\n",errur));
#
# Plot first derivative
  plot(r,ur,xlab="r",ylab="ur(r)");
   lines(r,ur,type="l",lwd=2);
  points(r,urs,pch="o",lwd=2);
#
# Second derivative (direct)
  urr=rep(0,n);
  for(ir in 1:n){
    urr[ir]=(-2*10*r[ir])^2*exp(-10*r[ir]^2)+
            -(2*10)*exp(-10*r[ir]^2);
  }
#
# urrs
  tabler=splinefun(r,us);
  urrs=tabler(r,deriv=2);
#
# Error norm
  errurr=0;
  for(ir in 1:n){
    errurr=errurr+(urrs[ir]-urr[ir])^2;
  }
  errurr=sqrt(errurr/n);
  cat(sprintf("\n errurr = %10.3e\n",errurr));
#
# Plot second derivative
  plot(r,urr,xlab="r",ylab="urr(r)");
   lines(r,urr,type="l",lwd=2);
  points(r,urrs,pch="o",lwd=2);
#
# Second derivative (stagewise)
  urr=rep(0,n);
  for(ir in 1:n){
    urr[ir]=(-2*10*r[ir])^2*exp(-10*r[ir]^2)+
            -(2*10)*exp(-10*r[ir]^2);
  }
```

(Continued)

```
                Listing A1.2 (Continued): Main program for moving
                                boundary, $r_u > 1$

#
# urs
  tabler=splinefun(r,us);
  urs=tabler(r,deriv=1);
#
# urrs
  tabler=splinefun(r,urs);
  urrs=tabler(r,deriv=1);
#
# Error norm
  errurr=0;
  for(ir in 1:n){
    errurr=errurr+(urrs[ir]-urr[ir])^2;
  }
  errurr=sqrt(errurr/n);
  cat(sprintf("\n errurr = %10.3e\n",errurr));
#
# Plot second derivative
  plot(r,urr,xlab="r",ylab="urr(r)");
   lines(r,urr,type="l",lwd=2);
  points(r,urrs,pch="o",lwd=2);
```

This routine is similar to the routine in Listing A1.1, so only the differences are noted.

- The r interval is extended by 25%

```
    #
    # us
      tabler=splinefun(r,u);
      ru=1.25;
      r=seq(from=rl,to=ru,by=(ru-rl)/(n-1));
      us=tabler(r,deriv=0);
      for(ir in 1:n){
        u[ir]=exp(-10*r[ir]^2);
      }
```

The r interval is redefined with this extension.

```
    r=seq(from=rl,to=ru,by=(ru-rl)/(n-1));
```

The calculation of u_s = us is then based on this extended interval.

```
    us=tabler(r,deriv=0);
```

The exact function $u(r)$ is also computed over the extended interval for the calculation of the norm of eq. (A1.2) and simultaneous plotting with $u_s(r)$.

```
u[ir]=exp(-10*r[ir]^2);
```

- Similarly, the first derivative $\dfrac{du_s(t)}{dr}$ is computed over the extended domain, then used in the norm of eq. (A1.4) and plotted simultaneously with $\dfrac{du(t)}{dr}$ of eq. (A1.3).

```
#
# urs
  tabler=splinefun(r,us);
  urs=tabler(r,deriv=1);
```

- The second derivative $\dfrac{d^2u_s(t)}{dr^2}$ is computed directly over the extended domain, then used in the norm of eq. (A1.6) and plotted simultaneously with $\dfrac{d^2u(t)}{dr^2}$ of eq. (A1.5).

```
#
# urrs
  tabler=splinefun(r,us);
  urrs=tabler(r,deriv=2);
```

- The second derivative $\dfrac{d^2u_s(t)}{dr^2}$ is computed successively over the extended domain, then used in the norm of eq. (A1.6) and plotted simultaneously with $\dfrac{d^2u(t)}{dr^2}$ of eq. (A1.5).

```
#
# urs
  tabler=splinefun(r,us);
  urs=tabler(r,deriv=1);
#
# urrs
  tabler=splinefun(r,urs);
  urrs=tabler(r,deriv=1);
```

For $r_l = 0 \le r \le r_u = 1$, the spline approximations correspond to interpolation and for $r_u = 1 \le r \le r_u = 1.25$, they correspond to extrapolation, which generally is less reliable (accurate) than interpolation.

The output from the routine of Listing A1.2 follows, first for $n = 21$, then $n = 41$.

The errors are generally greater than in Table A1.1 due to the extrapolation for $r_u > 1$, particularly for the error norm us = 7.983e-04.

The graphical output is in Figures A1.3-1 through A1.3-4.

When comparing the extrapolation errors for the first derivative (Figure A1.3-2) and the second derivative (Figures A1.3-3 and A1.3-4), the difference in the vertical scales should be noted. Generally, the second derivative errors are greater.

The reduction in the error for $n = 21$ changed to $n = 41$ is indicated by a comparison of Tables A1.3 and A1.4.

As a concluding case, n=41, ru=1.5 is considered for the extension of the grid to $0 \leq r \leq 1.5$. The numerical output is in Table A1.5.

As expected, the errors are larger for $r_u = 1.5$ (Table A1.5) than for $r_u = 1.25$ (Table A1.4).

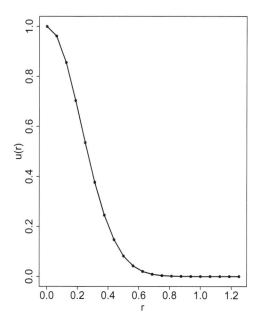

FIGURE A1.3-1

Comparison of moving boundary exact and spline solutions: ru=1.25, n=21.

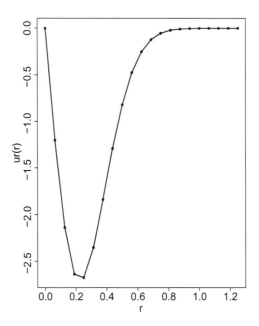

FIGURE A1.3-2

Comparison of moving boundary exact and spline first derivatives: ru=1.25, n=21.

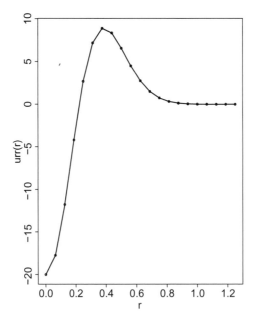

FIGURE A1.3-3

Comparison of moving boundary exact and spline second derivatives: Direct differentiation, ru=1.25, n=21.

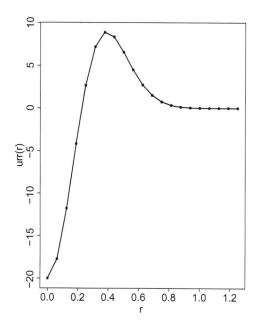

FIGURE A1.3-4
Comparison of moving boundary exact and spline second derivatives: Successive (stagewise) differentiation, ru=1.25, n=21.

TABLE A1.3
Numerical output, moving
boundary, n=21

erru =	7.983e−04
errur =	1.407e−02
errurr =	6.817e−01
errurr =	5.502e−01

TABLE A1.4
Numerical output, moving
boundary, n=41

erru =	2.606e-04
errur =	3.734e-03
errurr =	1.549e-01
errurr =	1.144e-01

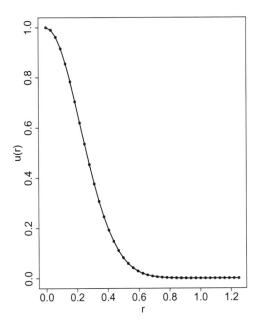

FIGURE A1.4-1
Comparison of moving boundary exact and spline solutions: ru=1.25, n=41.

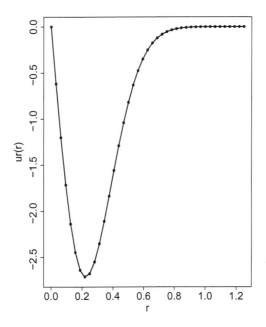

FIGURE A1.4-2
Comparison of moving boundary exact and spline first derivatives: `ru=1.25`, `n=41`.

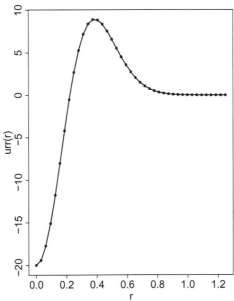

FIGURE A1.4-3
Comparison of moving boundary exact and spline second derivatives: Direct differentiation, `ru=1.25`, `n=41`.

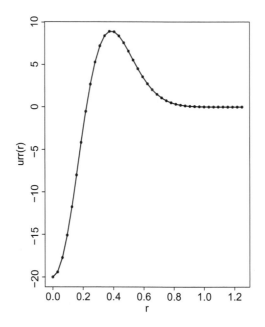

FIGURE A1.4-4

Comparison of moving boundary exact and spline second derivatives: Successive (stagewise) differentiation, ru=1.25, n=41.

TABLE A1.5

Numerical output, moving boundary, n=41, ru=1.5

erru =	2.614e-03
errur =	1.842e-02
errurr =	2.321e-01
errurr =	1.836e-01

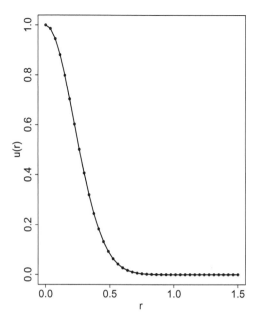

FIGURE A1.5-1
Comparison of moving boundary exact and spline solutions: ru=1.5, n=41.

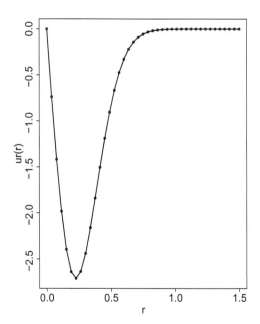

FIGURE A1.5-2
Comparison of moving boundary exact and spline first derivatives: ru=1.5, n=41.

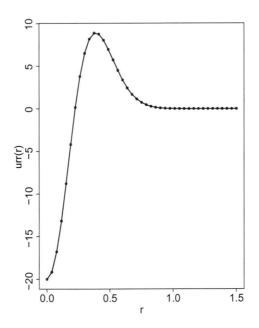

FIGURE A1.5-3
Comparison of moving boundary exact and spline second derivatives: Direct differentiation, `ru=1.5`, `n=41`.

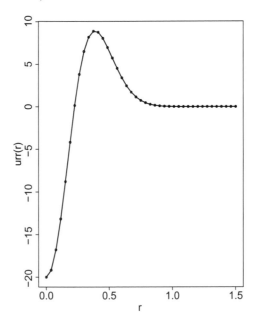

FIGURE A1.5-4
Comparison of moving boundary exact and spline second derivatives: Successive (stagewise) differentiation, `ru=1.5`, `n=41`.

The final conclusion is that attention should be given to the extrapolation region when using the moving boundary algorithm discussed in Chapters 1 through 3. Since an analytical solution for computing exact errors is generally not available (as it is in the example of this appendix), assessment of the PDE solution accuracy without the use of an analytical solution is required.

This can possibly be done by changing the number of grid points and observing the effect on the numerical solution, termed h refinement. This presumes that the accuracy of the extrapolated solution increases with increasing numbers of grid points. A second approach would be to use higher order splines than those in `splinefun`, termed p refinement. The objective of this form of error analysis is to achieve an acceptable level of spatial convergence of the numerical PDE solution.

In summary, some experimentation with the numerical solution to assess the accuracy of the moving boundary PDE solution will be required (generally by h refinement) in developing a moving boundary PDE application. An important advantage of splines is the redefinition of the spatial grid as the moving boundary PDE solution evolves. Generally, this cannot be accomplished as easily with conventional finite difference, finite volume, or finite element approximations that are intended for fixed grid PDE analysis.

Appendix A2

Test of Moving Boundary Algorithm Spline Formulation

In this appendix, a conventional method of lines (MOL) analysis is implemented in which one call to the ordinary differential equation (ODE) integrator **ode** gives a partial differential equation (PDE) solution for the complete interval in t ($0 \leq t \leq 0.2$). This solution is then compared with the solution from the moving boundary PDE algorithm discussed in Chapter 2 for the special case of a fixed boundary (Listing 2.1, **ncase=1**) to provide a test of the moving boundary algorithm for this special case.

A2.1 Main Program

The main program for the special case test follows.

Listing A2.1: Main program for the special case test with splines

```
#
# Delete previous workspaces
  rm(list=ls(all=TRUE))
#
# Access ODE integrator
  library("deSolve");
#
# Access functions for numerical solution
  setwd("f:/mbpde/appA2");
  source("pde_1a.R");
#
# Grid (in r)
  n=21;rl=0;ru=1;
  r=seq(from=rl,to=ru,by=(ru-rl)/(n-1));
```

(Continued)

Listing A2.1 (Continued): Main program for the special case
test with splines

```
#
# Parameters
  D=1;
  km=0;
  ua=1;
#
# Independent variable for ODE integration
  nout=6;t0=0;tf=0.2;
  tout=seq(from=t0,to=tf,by=(tf-t0)/(nout-1));
#
# Initial condition
  u0=exp(-10*r^2);
  ncall=0;
#
# ODE integration
  out=ode(func=pde_1a,y=u0,times=tout);
  nrow(out)
  ncol(out)
#
# Array for display
  u=matrix(0,nrow=n,ncol=nout);
  for(it in 1:nout){
  for(ir in 1:n){
    u[ir,it]=out[it,ir+1];
  }
  }
#
# Numerical solution
  for(it in 1:nout){
  cat(sprintf("\n\n        t       r      u(r,t)"));
  iv=seq(from=1,to=n,by=5);
  for(ir in iv){
    cat(sprintf("\n %6.2f%6.2f%10.4f",
      tout[it],r[ir],u[ir,it]));
  }
  }
  matplot(r,u,type="l",lwd=2,col="black",lty=1,
    xlab="x",ylab="u(r,t)");
#
# Calls to ODE routine
  cat(sprintf("\n\n  ncall = %3d\n",ncall));
```

We can note the following details about Listing A2.1.

- Previous workspaces are deleted.

```
#
# Delete previous workspaces
  rm(list=ls(all=TRUE))
```

- The R ODE integrator library deSolve is accessed. Then the directory with the files for the solution of eqs. (1.2) through (1.4) is designated. Note that setwd (set working directory) uses / rather than the usual \.

```
#
# Access ODE integrator
  library("deSolve");
#
# Access functions for numerical solution
  setwd("f:/mbpde/appA2");
  source("pde_1a.R");
```

pde_1a.R is the routine for the method of lines (MOL) approximation of PDE (1.2).

- A spatial grid of 21 points is defined for $x_l = 0 \le x \le x_u = 1$, so that x=0,0.05,...,1.

```
#
# Grid (in r)
  n=21;rl=0;ru=1;
  r=seq(from=rl,to=ru,by=(ru-rl)/(n-1));
```

- The model parameters are defined numerically.

```
#
# Parameters
  D=1;
  km=0;
  ua=1;
```

- An interval in t of 6 points is defined for $0 \le t \le 0.2$ so that tout=0,0.04,...,0.2.

```
#
# Independent variable for ODE integration
  nout=6;t0=0;tf=0.2;
  tout=seq(from=t0,to=tf,by=(tf-t0)/(nout-1));
```

- The initial condition (IC) function of eq. (1.3) is defined as a Gaussian function centered at $r = 0$.

```
#
# Initial condition
  u0=exp(-10*r^2);
  ncall=0;
```

The statement for u0 is based on the R vector facility since r is a vector. The counter for the calls to the ODE/MOL routine pde_1a is also initialized.

- The integration of the $n = 21$ ODEs is by this single call to ode for the interval $t_0 = 0 \le t \le t_f = 0.2$ which is a conventional MOL analysis (in comparison to the series of calls to ode in Listing 2.1). The inputs to ode are the ODE function, pde_1a, the IC vector u0, and the vector of output values of t, tout. The length of u0 (21) informs ode how many ODEs are to be integrated. func,y,times are reserved names.

```
#
# ODE integration
  out=ode(func=pde_1a,y=u0,times=tout);
  nrow(out)
  ncol(out)
```

The numerical solution to the ODEs is returned in matrix out. In this case, out has the dimensions $nout \times (nx + 1) = 6 \times 21 + 1 = 22$, which are confirmed by the output from nrow(out),ncol(out) (included in the numerical output considered subsequently).

The offset $21 + 1$ is required since the first element of each column has the output t (also in tout), and the $2, ..., n + 1 = 2, ..., 22$ column elements have the 21 ODE solutions.

- The solutions of the 21 ODEs returned in out by ode are placed in array u.

```
#
# Array for display
  u=matrix(0,nrow=n,ncol=nout);
  for(it in 1:nout){
  for(ir in 1:n){
    u[ir,it]=out[it,ir+1];
  }
  }
```

Again, the offset ir+1 is required since the first element of each column of out has the value of t.

- $u(r, t)$ is displayed as a function of r and t, with every fifth value of r from by=5.

```
#
# Numerical solution
  for(it in 1:nout){
  cat(sprintf("\n\n      t      r      u(r,t)"));
  iv=seq(from=1,to=n,by=5);
  for(ir in iv){
    cat(sprintf("\n %6.2f%6.2f%10.4f",
      tout[it],r[ir],u[ir,it]));
  }
  }
```

- $u(r, t)$ is plotted as a function of r with t as a parameter.

```
matplot(r,u,type="l",lwd=2,col="black",lty=1,
  xlab="x",ylab="u(r,t)");
```

- The number of calls to **pde_1a** is displayed at the end of the solution.

```
#
# Calls to ODE routine
  cat(sprintf("\n\n  ncall = %3d\n",ncall));
```

The ODE/MOL routine **pde_1a** called by **ode** is considered next.

A2.2 ODE/MOL Routine

Listing A2.2: ODE/MOL routine for eqs. (1.2) through (1.4)

```
pde_1a=function(t,u,parms){
#
# Function pde_1a computes the t derivative
# vector of u(r,t)
#
# ur
  tabler=splinefun(r,u);
  ur=tabler(r,deriv=1);
```

(Continued)

**Listing A2.2 (Continued): ODE/MOL routine for eqs. (1.2)
through (1.4)**

```
#
# BCs
  ur[1]=0;ur[n]=(km/D)*(ua-u[n]);
#
# urr
  tabler=splinefun(r,ur);
  urr=tabler(r,deriv=1);
#
# PDEs
  ut=rep(0,n);
  for(i in 1:n){
    if(i==1){
      ut[i]=2*D*urr[i];
    }else{
      ut[i]=D*(urr[i]+(1/r[i])*ur[i]);
    }
  }
#
# Increment calls to pde_1a
  ncall <<- ncall+1;
#
# Return derivative vector
  return(list(c(ut)));
  }
```

---

pde_1a is the same as in Listing (2.2) so the discussion is not repeated here.

This completes the programming of eqs. (1.2) through (1.4). The output from the routines of Listings A2.1 and A2.2 is discussed next.

A2.3   Model Output

The numerical output from the main program and ODE/MOL routines in Listings A2.1 and A2.2 follows (Table A2.1).

TABLE A2.1

Numerical output for Eqs. (1.2)
through (1.4), mf=0, n=21

[1] 6

[1] 22

t	r	u(r,t)
0.00	0.00	1.0000
0.00	0.25	0.5353
0.00	0.50	0.0821
0.00	0.75	0.0036
0.00	1.00	0.0000

t	r	u(r,t)
0.04	0.00	0.3840
0.04	0.25	0.3025
0.04	0.50	0.1471
0.04	0.75	0.0456
0.04	1.00	0.0172

t	r	u(r,t)
0.08	0.00	0.2378
0.08	0.25	0.2057
0.08	0.50	0.1335
0.08	0.75	0.0712
0.08	1.00	0.0483

t	r	u(r,t)
0.12	0.00	0.1737
0.12	0.25	0.1578
0.12	0.50	0.1196
0.12	0.75	0.0843
0.12	1.00	0.0703

t	r	u(r,t)
0.16	0.00	0.1404
0.16	0.25	0.1321
0.16	0.50	0.1110
0.16	0.75	0.0913
0.16	1.00	0.0832

(Continued)

TABLE A2.1 (*Continued*)
Numerical output for Eqs. (1.2)
through (1.4), `mf=0`, `n=21`

t	r	u(r,t)
0.20	0.00	0.1222
0.20	0.25	0.1178
0.20	0.50	0.1061
0.20	0.75	0.0952
0.20	1.00	0.0904

`ncall = 374`

This solution is essentially identical to the solution in Table 3.1 which confirms that the MBPDE algorithm of Chapters 1 and 2 with incremental calls to ode (`ncase=1` in Listing 2.1) gives the same solution as the MOL solution with one call to ode (**Listing A2.1**) for the case of a fixed boundary. In other words, the MBPDE algorithm reduces to the conventional MOL algorithm for a fixed boundary as a special case.

In Appendix A3, the preceding test is repeated with finite differences (FDs) used in **pde_1a** in place of splines.

Appendix A3

Test of Moving Boundary Algorithm Finite Difference Formulation

In this appendix, a conventional method of lines (MOL) analysis is implemented which parallels the analysis in Appendix A2, but with finite differences (FDs) used in place of splines. Since the coding is similar, only the differences from Listings A2.1 and A2.2 will be considered.

A3.1 Main Program

The main program for the special case test follows.

Listing A3.1: Main program for the special case test with finite differences

```
#
# Delete previous workspaces
  rm(list=ls(all=TRUE))
#
# Access ODE integrator
  library("deSolve");
#
# Access functions for numerical solution
  setwd("f:/mbpde/appA3");
  source("pde_1a.R");
  source("dss004.R");
  source("dss044.R");
#
# Grid (in r)
  n=21;rl=0;ru=1;
  r=seq(from=rl,to=ru,by=(ru-rl)/(n-1));
#
# Parameters
  D=1;
```

(Continued)

Listing A3.1 (Continued): Main program for the special case test with finite differences

```
  km=0;
  ua=1;
  ndss=1;
#
# Independent variable for ODE integration
  nout=6;t0=0;tf=0.2;
  tout=seq(from=t0,to=tf,by=(tf-t0)/(nout-1));
#
# Initial condition
  u0=exp(-10*r^2);
  ncall=0;
#
# ODE integration
  out=ode(func=pde_1a,y=u0,times=tout);
  nrow(out)
  ncol(out)
#
# Array for display
  u=matrix(0,nrow=n,ncol=nout);
  for(it in 1:nout){
  for(ir in 1:n){
    u[ir,it]=out[it,ir+1];
  }
  }
#
# Numerical solution
  for(it in 1:nout){
  cat(sprintf("\n\n        t       r      u(r,t)"));
  iv=seq(from=1,to=n,by=5);
  for(ir in iv){
    cat(sprintf("\n %6.2f%6.2f%10.4f",
      tout[it],r[ir],u[ir,it]));
  }
  }
  matplot(r,u,type="l",lwd=2,col="black",lty=1,
    xlab="x",ylab="u(r,t)");
#
# Calls to ODE routine
  cat(sprintf("\n\n  ncall = %3d\n",ncall));
```

We can note the following details about Listing A3.1.

- The R ordinary differential equation (ODE) integrator library `deSolve` is accessed. Then the directory with the files for the solution of eqs. (1.2) through (1.4) is designated. Note that `setwd` (set working directory) uses / rather than the usual \.

```
#
# Access ODE integrator
  library("deSolve");
#
# Access functions for numerical solution
  setwd("f:/mbpde/appA3");
  source("pde_1a.R");
```

`pde_1a.R` is the routine for the method of lines (MOL) approximation of PDE (1.2). `dss004`, `dss044` are library routines with FDs for first and second order spatial derivatives [1].

- The model parameters are defined numerically.

```
#
# Parameters
  D=1;
  km=0;
  ua=1;
  ndss=1;
```

`ndss` is a parameter to select a FD in `pde_1a` as explained subsequently.

Otherwise, Listings A2.1 and A3.1 are the same. The ODE/MOL routine `pde_1a` follows.

A3.2   ODE/MOL Routine

Listing A3.2: ODE/MOL routine for eqs. (1.2) through (1.4) with finite differences

```
pde_1a=function(t,u,parms){
#
# Function pde_1a computes the t derivative
# vector of u(r,t)
#
```

(Continued)

Listing A3.2 (Continued): ODE/MOL routine for eqs. (1.2)
                  through (1.4) with finite differences

```
# ur
  ur=dss004(rl,ru,n,u);
#
# BCs
  ur[1]=0;ur[n]=(km/D)*(ua-u[n]);
#
# urr
  if(ndss==1){urr=dss004(rl,ru,n,ur);}
  if(ndss==2){nl=2;nu=2;
    urr=dss044(rl,ru,n,u,ur,nl,nu);}
#
# PDEs
  ut=rep(0,n);
  for(i in 1:n){
    if(i==1){
      ut[i]=2*D*urr[i];
    }else{
      ut[i]=D*(urr[i]+(1/r[i])*ur[i]);
    }
  }
#
# Increment calls to pde_1a
  ncall <<- ncall+1;
#
# Return derivative vector
  return(list(c(ut)));
  }
```

We can note the following details about Listing A3.2.

- The function is defined.

  ```
    pde_1a=function(t,u,parms){
  #
  # Function pde_1a computes the t derivative
  # vector of u(r,t)
  ```

  Additional details about the input and output arguments are given
  in the discussion of Listing 2.2.

- The first partial derivative $\dfrac{\partial u}{\partial r}$ is computed with dss004 [1].

```
#
# ur
   ur=dss004(rl,ru,n,u);
```

- Boundary conditions (BCs) (1.4) are programmed.

```
#
# BCs
   ur[1]=0;ur[n]=(km/D)*(ua-u[n]);
```

Subscripts 1,n correspond to $r = r_l, r_u$, respectively.

- The second partial derivative $\dfrac{\partial^2 u}{\partial r^2}$ is computed.

```
#
# urr
   if(ndss==1){urr=dss004(rl,ru,n,ur);}
   if(ndss==2){nl=2;nu=2;
      urr=dss044(rl,ru,n,u,ur,nl,nu);}
```

For ndss=1 (set in the main program of Listing A3.1), successive (stagewise) differentiation is used to calculate the second derivative, that is, the first derivative ur is differentiated. For ndss=2, u is differentiated directly to the second derivative, urr. nl=2, nu=2 designate Neumann BCs (ur[1], ur[n] are specified).

- Equation (1.2) is programmed.

```
#
# PDEs
   ut=rep(0,n);
   for(i in 1:n){
      if(i==1){
         ut[i]=2*D*urr[i];
      }else{
         ut[i]=D*(urr[i]+(1/r[i])*ur[i]);
      }
   }
```

The derivative $\dfrac{\partial u}{\partial t}$ is placed in vector ut. For i=1 corresponding to $r_l = 0$, the term $\dfrac{1}{r}\dfrac{\partial u}{\partial r}$ is indeterminant $(0/0)$ and is resolved with l'Hospital's rule [1]. That is, the radial group in eq. (1.2) at $r = 0$ is

$$\frac{\partial^2 u}{\partial r^2} + \frac{1}{r}\frac{\partial u}{\partial r} = 2\frac{\partial^2 u}{\partial r^2},$$

and is programmed as `2*D*urr[i]`. For `i>1`, the programming follows directly from eq. (1.2), which demonstrates a principal feature of the MOL (the close correspondence of the PDE and MOL programming).

- The counter for the calls to `pde_1a` is incremented and returned to the main program of Listing 2.1 with `<<-`.

```
#
# Increment calls to pde_1a
   ncall <<- ncall+1;
```

- `ut` is returned to `ODE` as a list (required by `ODE`). `c` is the R vector utility.

```
#
# Return derivative vector
   return(list(c(ut)));
   }
```

The final `}` concludes `pde_1a`.

This completes the programming of eqs. (1.2) through (1.4). The output from the routines of Listings A3.1 and A3.2 is discussed next.

A3.3   Model Output

The numerical output from the main program and ODE/MOL routines in Listings A3.1 and A3.2 follows (Table A3.1).

This solution is close to the solution in Table 3.1, e.g., at $t = 0.2$,

```
Table 3.1 (Listings 2.1 and 2.2)

    t      r      u(r,t)
  0.20   0.00    0.1222
  0.20   0.25    0.1178
  0.20   0.50    0.1061
  0.20   0.75    0.0952
  0.20   1.00    0.0904

  ncall = 573


Table A3.1 (Listings A3.1 and A3.2)

    t      r      u(r,t)
  0.20   0.00    0.1229
```

```
0.20   0.25     0.1177
0.20   0.50     0.1062
0.20   0.75     0.0951
0.20   1.00     0.0911

ncall = 311
```

which again confirms that the moving boundary partial differential equation (MBPDE) algorithm of Chapters 1 and 2 with incremental calls to ode (ncase=1 in Listing 3.1) gives the same solution as the conventional MOL solution with one call to ode (Listing A3.1) for the case of a fixed boundary.

Finally, the output with ndss=2 in Listing A3.1 follows (Table A3.2).

This solution is close to the solution in Table 3.1, e.g., at $t = 0.2$,

Table 3.1 (Listings 2.1 and 2.2)

```
  t       r      u(r,t)
0.20   0.00     0.1222
0.20   0.25     0.1178
0.20   0.50     0.1061
0.20   0.75     0.0952
0.20   1.00     0.0904

ncall = 573
```

Table A3.1 (Listings A3.1 and A3.2, ndss=1)

```
  t       r      u(r,t)
0.20   0.00     0.1229
0.20   0.25     0.1177
0.20   0.50     0.1062
0.20   0.75     0.0951
0.20   1.00     0.0911

ncall = 311
```

Table A3.2 (Listings A3.1 and A3.2, ndss=2)

```
  t       r      u(r,t)
0.20   0.00     0.1227
0.20   0.25     0.1178
0.20   0.50     0.1062
0.20   0.75     0.0951
0.20   1.00     0.0909

ncall = 423
```

TABLE A3.1

Numerical output for Eqs. (1.2) through (1.4), `mf=0`, `n=21`, `ndss=1`

[1] 6

[1] 22

t	r	u(r,t)
0.00	0.00	1.0000
0.00	0.25	0.5353
0.00	0.50	0.0821
0.00	0.75	0.0036
0.00	1.00	0.0000

t	r	u(r,t)
0.04	0.00	0.3849
0.04	0.25	0.3024
0.04	0.50	0.1472
0.04	0.75	0.0455
0.04	1.00	0.0176

t	r	u(r,t)
0.08	0.00	0.2385
0.08	0.25	0.2056
0.08	0.50	0.1336
0.08	0.75	0.0711
0.08	1.00	0.0489

t	r	u(r,t)
0.12	0.00	0.1744
0.12	0.25	0.1577
0.12	0.50	0.1197
0.12	0.75	0.0842
0.12	1.00	0.0709

t	r	u(r,t)
0.16	0.00	0.1411
0.16	0.25	0.1319
0.16	0.50	0.1111
0.16	0.75	0.0912
0.16	1.00	0.0838

(Continued)

TABLE A3.1 (*Continued*)

Numerical output for Eqs. (1.2) through (1.4), mf=0, n=21, ndss=1

t	r	u(r,t)
0.20	0.00	0.1229
0.20	0.25	0.1177
0.20	0.50	0.1062
0.20	0.75	0.0951
0.20	1.00	0.0911

ncall = 311

TABLE A3.2

Numerical output for Eqs. (1.2) through (1.4), mf=0, n=21, ndss=2

t	r	u(r,t)
0.00	0.00	1.0000
0.00	0.25	0.5353
0.00	0.50	0.0821
0.00	0.75	0.0036
0.00	1.00	0.0000
t	r	u(r,t)
0.04	0.00	0.3846
0.04	0.25	0.3024
0.04	0.50	0.1472
0.04	0.75	0.0455
0.04	1.00	0.0175
t	r	u(r,t)
0.08	0.00	0.2383
0.08	0.25	0.2057
0.08	0.50	0.1335
0.08	0.75	0.0712
0.08	1.00	0.0487
t	r	u(r,t)
0.12	0.00	0.1742
0.12	0.25	0.1578
0.12	0.50	0.1196

(*Continued*)

TABLE A3.2 (*Continued*)
Numerical output for Eqs. (1.2) through
(1.4), `mf=0`, `n=21`, `ndss=2`

t	r	u(r,t)
0.12	0.75	0.0842
0.12	1.00	0.0707
t	r	u(r,t)
0.16	0.00	0.1409
0.16	0.25	0.1320
0.16	0.50	0.1111
0.16	0.75	0.0912
0.16	1.00	0.0836
t	r	u(r,t)
0.20	0.00	0.1227
0.20	0.25	0.1178
0.20	0.50	0.1062
0.20	0.75	0.0951
0.20	1.00	0.0909

`ncall = 423`

In summary, the MBPDE algorithm of Chapters 1 and 2 has a special case solution for a fixed boundary that is in agreement with the solution of a conventional MOL analysis based on splines and FDs. This special case test provides confirmation that the MBPDE algorithm reduces to the fixed boundary MOL algorithm.

Reference

1. Schiesser, W.E. (2014), *Differential Equation Analysis in Biomedical Science and Engineering: Partial Differential Equation Applications in R*, John Wiley & Sons, Hoboken, NJ.

Index

Printed and bound by CPI Group (UK) Ltd, Croydon, CR0 4YY

17/10/2024

01775681-0020